Anatomy and Physiology
A Self-Instructional Course

Anatomy and Physiology
A Self-Instructional Course

5. The Urinary System and The Digestive System

Written and designed by
Cambridge Communication Limited

Medical adviser

Bryan Broom MB BS(Lond)
General Practitioner
Beit Memorial Research Fellow
Middlesex Hospital Medical Research School

SECOND EDITION

Churchill Livingstone ⊞
EDINBURGH LONDON MELBOURNE AND NEW YORK 1985

CHURCHILL LIVINGSTONE
Medical Division of Longman Group Limited

Distributed in the United States of America by
Churchill Livingstone Inc., 650 Avenue of the
Americas, New York, N.Y. 10011, and by
associated companies, branches and
representatives throughout the world.

First edition 1977
Second edition 1985
  Reprinted 1991, 1993

ISBN 0-443-03209-2

British Library Cataloguing in Publication Data
A catalogue record for this book is available from
the British Library

Library of Congress Cataloging in Publication Data
Anatomy and physiology.
  Rev. ed. of: anatomy and physiology /
Ralph Rickards, David F. Chapman. 1977.
  Contents: 1. The human body and the reproductive
system — 2. The endocrine glands and the nervous
system — 3. The locomotor system and the special
senses — [etc.]
  1. Human physiology — Programmed instruction.
2. Anatomy, Human — Programmed instruction.
I. Broom, Bryan. II. Rickards, Ralph. Anatomy and
physiology. III. Cambridge Communication Limited.
QP34.5.A47 1984   612   84-4977

The
publisher's
policy is to use
paper manufactured
from sustainable forests

Printed in Hong Kong
WC/03

Contents

Contents

The Urinary System

1. Introduction

The body of an average man, weighing 70 kg, contains about 42 litres of water.

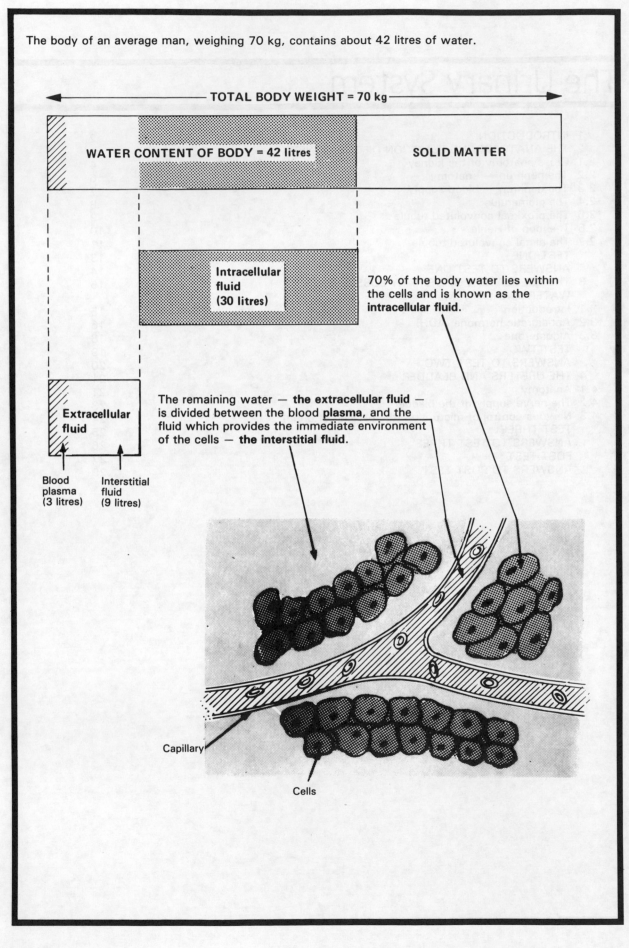

◄─────── **TOTAL BODY WEIGHT = 70 kg** ───────►

WATER CONTENT OF BODY = 42 litres   **SOLID MATTER**

Intracellular fluid (30 litres)

70% of the body water lies within the cells and is known as the **intracellular fluid**.

Extracellular fluid

The remaining water — **the extracellular fluid** — is divided between the blood **plasma**, and the fluid which provides the immediate environment of the cells — **the interstitial fluid**.

Blood plasma (3 litres)

Interstitial fluid (9 litres)

Capillary

Cells

Plasma and interstitial fluid have a similar composition. They are aqueous solutions of inorganic ions (mainly sodium and chloride ions, with much lesser amounts of potassium, calcium, magnesium, bicarbonate and phosphate ions). They also contain nutrients to the cells (amino acids, fats and glucose), waste products from the cells, and traces of hormones, vitamins and enzymes. Plasma (but not interstitial fluid) contains some protein molecules.

The kidneys:

1. **filter** nitrogenous wastes (mainly as urea) and other toxins from the blood;

2. **control** the loss of water and electolytes in the urine, thus maintaining the correct balance of these substances in the body.

Controlling the composition and volume of the extracellular fluid, that is maintaining the closely regulated environment required by the delicately structured cells if they are to function properly, is carried out by the kidneys.

2. The anatomy and function of the kidney

2.1. Gross anatomy of the kidney

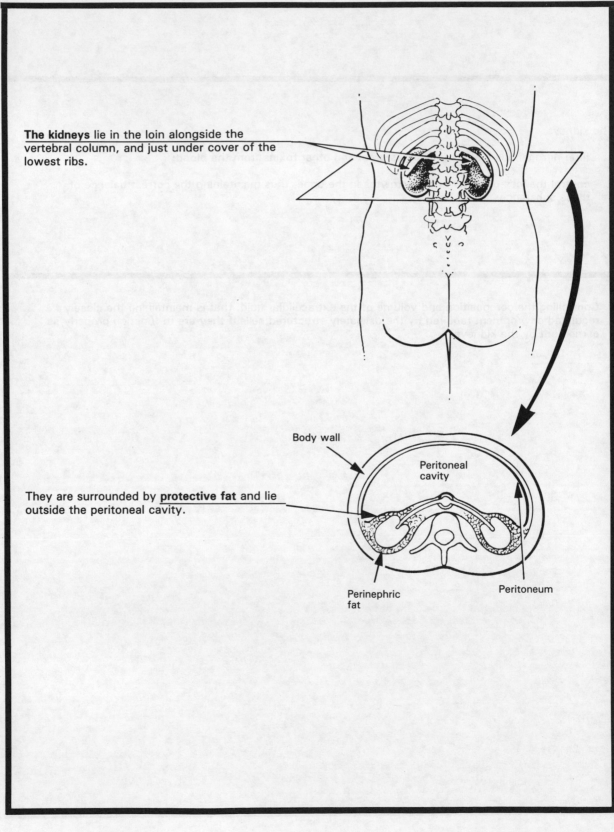

The kidneys lie in the loin alongside the vertebral column, and just under cover of the lowest ribs.

They are surrounded by **protective fat** and lie outside the peritoneal cavity.

Body wall

Peritoneal cavity

Peritoneum

Perinephric fat

A quarter of the output of the heart (1300 ml of blood per minute) circulates through the kidneys.

The blood enters the hilum of each kidney via the **renal artery**. _____

Some is filtered off, and processed to form urine.

The blood leaves by the **renal vein**. _____

The urine collects in the **pelvis** of the kidney, leaves via the **ureter**, _____

is stored in the **bladder** _____

and discharged through the **urethra**. _____

2.2. The nephron—anatomy

The kidney is composed of a mass of microscopic tubules called nephrons which filter the blood and control its composition. There are about a million **nephrons** in each kidney.

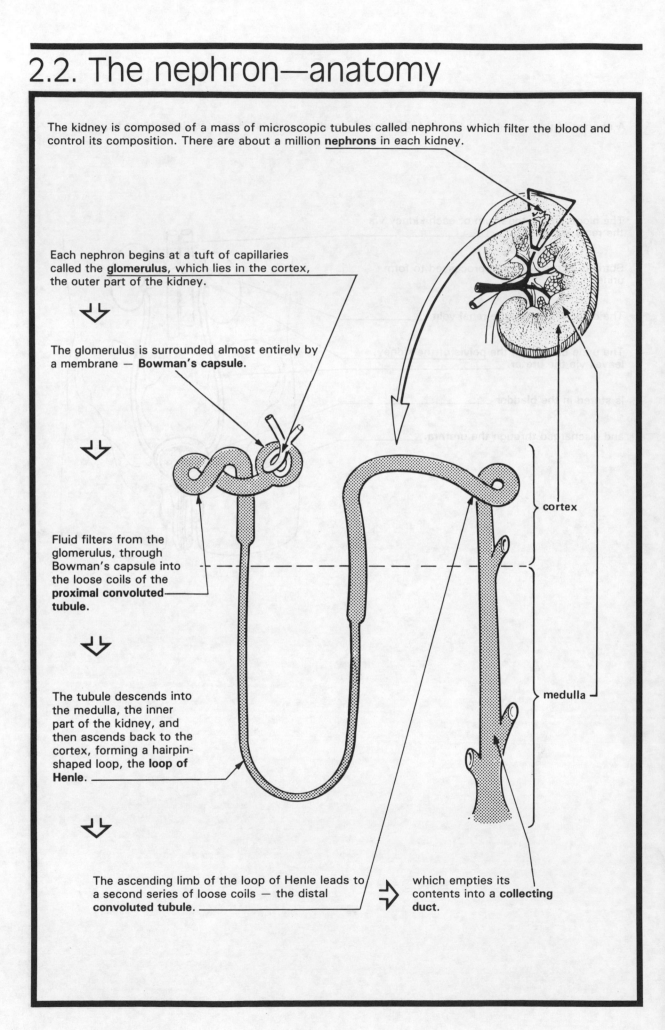

Each nephron begins at a tuft of capillaries called the **glomerulus**, which lies in the cortex, the outer part of the kidney.

The glomerulus is surrounded almost entirely by a membrane — **Bowman's capsule**.

Fluid filters from the glomerulus, through Bowman's capsule into the loose coils of the **proximal convoluted tubule**.

The tubule descends into the medulla, the inner part of the kidney, and then ascends back to the cortex, forming a hairpin-shaped loop, the **loop of Henle**.

The ascending limb of the loop of Henle leads to a second series of loose coils — the distal **convoluted tubule**.

which empties its contents into a **collecting duct**.

cortex

medulla

2.3. The nephron—blood supply

Blood enters the kidney through the renal artery which divides into branches which pass into the cortex.

The branches end at the glomeruli where they form tufts of very fine capillaries. All the blood entering the kidney first passes through the **glomeruli**.

The glomerular capillaries reunite to form a **single arteriole**, which then divides again into a **network of capillaries around the tubules** giving them a rich blood supply.

cortex

medulla

These capillaries finally reunite and drain into **tributaries of the renal vein**.

2.4. The glomerulus

1300 ml of blood enter the glomeruli every minute.

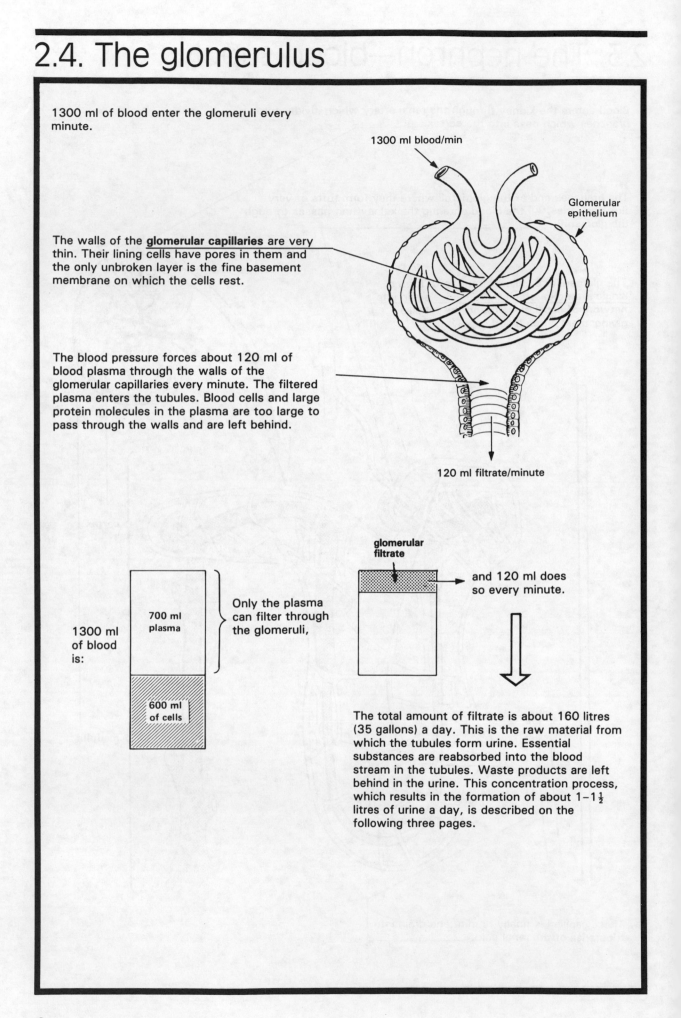

1300 ml blood/min

Glomerular epithelium

The walls of the **glomerular capillaries** are very thin. Their lining cells have pores in them and the only unbroken layer is the fine basement membrane on which the cells rest.

The blood pressure forces about 120 ml of blood plasma through the walls of the glomerular capillaries every minute. The filtered plasma enters the tubules. Blood cells and large protein molecules in the plasma are too large to pass through the walls and are left behind.

120 ml filtrate/minute

glomerular filtrate

and 120 ml does so every minute.

1300 ml of blood is:

700 ml plasma

600 ml of cells

Only the plasma can filter through the glomeruli,

The total amount of filtrate is about 160 litres (35 gallons) a day. This is the raw material from which the tubules form urine. Essential substances are reabsorbed into the blood stream in the tubules. Waste products are left behind in the urine. This concentration process, which results in the formation of about 1–1½ litres of urine a day, is described on the following three pages.

2.5. The proximal convoluted tubule

Most of the glomerular filtrate is absorbed back into the bloodstream through the **capillaries surrounding the proximal convoluted tubule.**

Glucose, amino acids, and potassium ions are completely reabsorbed here. Most of the filtered sodium ions and water are also reabsorbed here.

Creatinine, sulphates, most of the urea, and other waste products are not reabsorbed, but stay in the forming urine.

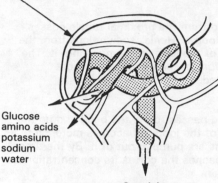

Glucose
amino acids
potassium
sodium
water

Creatinine
sulphates
urea

The absorption of sodium ions takes place by an active process in which the **cells lining the tubule take up the ions** and then pump them into the **bloodstream.**

Capillaries
surrounding
tubule

Sodium
ions

This transfer of sodium ions is from a dilute solution to a more concentrated solution and requires a great deal of energy.

About 1/8th of the energy used by a resting man is consumed by this 'sodium pump' mechanism of the renal tubules.

About 85% of the filtered sodium ions are pumped out of the proximal tubules back into the bloodstream. Chloride ions accompany the sodium ions, and about 85% of the filtered water is simultaneously reabsorbed by osmosis.

The remaining filtrate passes into the loop of Henle.

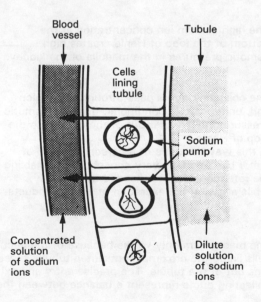

Blood
vessel

Tubule

Cells
lining
tubule

'Sodium
pump'

Concentrated
solution
of sodium
ions

Dilute
solution
of sodium
ions

2.6. The loop of Henle

The renal tubular apparatus works on the counter-current principle.

As the fluid from the proximal tubule passes down the **loop of Henle**, sodium ions from the other side of the loop are pumped into it. The fluid therefore becomes more and more concentrated.

As the fluid passes round the bend and up the other side of the loop, it becomes more dilute as sodium ions are pumped out of it. By the time the fluid reaches the cortex its concentration is normal again.

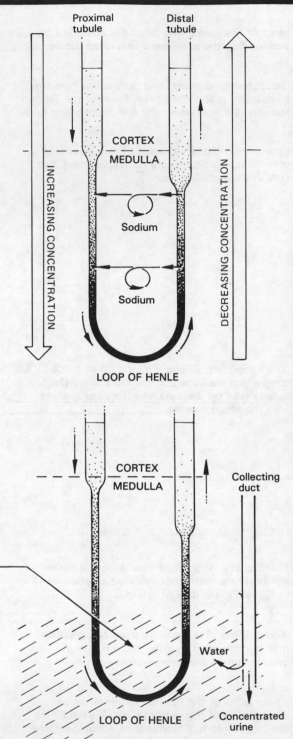

The high sodium ion concentration at the bottom of the loop of Henle creates high osmotic pressures in the medulla of the kidney.

The **collecting ducts** pass through this region and, under certain conditions, this high osmotic pressure draws water out of the ducts into the loop of Henle.
In this way the urine of a person deprived of water can be made very concentrated, enabling the retention of water needed by the body, while allowing the excretion of waste products.

The mechanism only works because of the peculiar nature of cells lining the loop of Henle. Unlike the cells lining the proximal convoluted tubule, these cells do not allow water to accompany the sodium ions out of the tubule. The precise entry and exit of salts and water to and from the tubules and collecting ducts represent a balance between the opposing forces exerted by:

1. the pressure and flow in the renal blood vessels;
2. the permeability of these vessels, the tubules and collecting ducts;
3. the hydrostatic pressure in the vessels, the tubules and collecting ducts;
4. factors affecting these variables, such as ADH and aldosterone, especially in diseased states.

2.7. The distal convoluted tubule

Fine adjustments to the composition of the urine are made at the distal convoluted tubule. Only about 15% of the glomerular filtrate (about 20 ml/minute) reaches the distal tubule, the remainder having already been absorbed in the proximal tubule.

At the **distal tubule** the following processes take place.

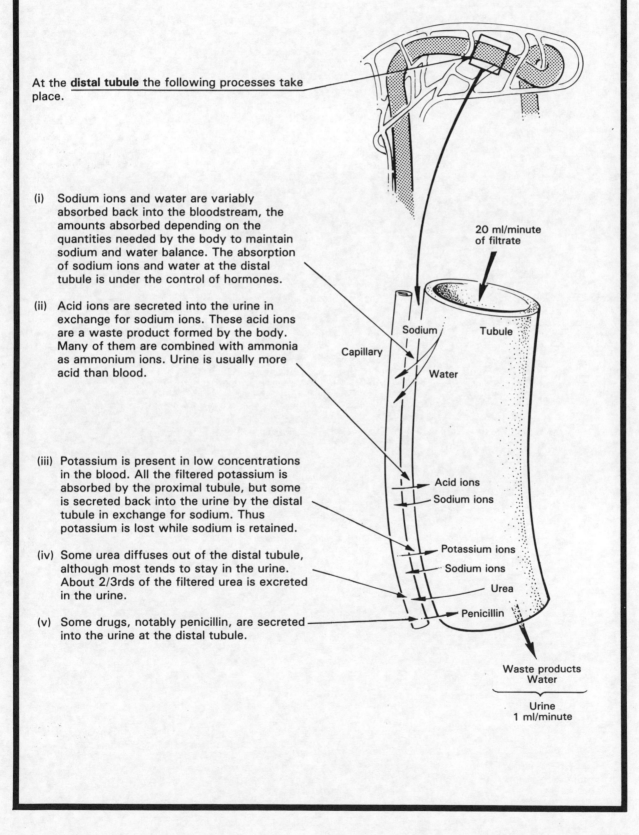

(i) Sodium ions and water are variably absorbed back into the bloodstream, the amounts absorbed depending on the quantities needed by the body to maintain sodium and water balance. The absorption of sodium ions and water at the distal tubule is under the control of hormones.

(ii) Acid ions are secreted into the urine in exchange for sodium ions. These acid ions are a waste product formed by the body. Many of them are combined with ammonia as ammonium ions. Urine is usually more acid than blood.

(iii) Potassium is present in low concentrations in the blood. All the filtered potassium is absorbed by the proximal tubule, but some is secreted back into the urine by the distal tubule in exchange for sodium. Thus potassium is lost while sodium is retained.

(iv) Some urea diffuses out of the distal tubule, although most tends to stay in the urine. About 2/3rds of the filtered urea is excreted in the urine.

(v) Some drugs, notably penicillin, are secreted into the urine at the distal tubule.

20 ml/minute of filtrate

Sodium

Tubule

Capillary

Water

Acid ions

Sodium ions

Potassium ions

Sodium ions

Urea

Penicillin

Waste products
Water

Urine
1 ml/minute

TEST ONE

1. **The urinary system can be represented functionally like this:**

Write down the names of the parts of the urinary system represented above by:

(a) _____ (c) _____ (e) _____

(b) _____ (d) _____ (f) _____

2. **Which of the following is the most important electrolyte (ion) in extracellular fluid?**

   (a) Potassium.   (b) Bicarbonate.   (c) Sodium.   (d) Glucose.   (e) Magnesium.   (f) Phosphate.

3. **Tick the appropriate brackets to show the location:**

	In the cortex of the kidney	In the medulla of the kidney
Bowman's capsule.	( )	( )
The distal convoluted tubule.	( )	( )
The proximal convoluted tubule.	( )	( )
The loop of Henle.	( )	( )
The glomerulus.	( )	( )

4. **What is the flow rate of blood per minute through the kidney?**

   (a) 200 ml   (b) 700 ml   (c) 1300 ml   (d) 5000 ml.

5. **What is the volume of fluid filtered by the glomeruli in one minute?**

   (a) 10 ml   (b) 120 ml   (c) 300 ml   (d) 700 ml.

6. **Which of the following can be secreted back into the distal tubule from the blood?**

   (a) Vitamins.   (c) Water.   (e) Glucose.
   (b) Potassium ions.   (d) Acid ions.

7. **Which of the descriptions listed on the right apply to the parts of the kidney listed on the left?**

   (a) The proximal tubule.   (i)   Creates high osmotic pressures in the medulla.
   (b) The loop of Henle.   (ii) Secretes ammonium ions into the tubule.
   (c) The distal tubule.   (iii) Reabsorbs 85% of the volume of the
                                           glomerular filtrates.

8. **Where there is insufficient intake of water urine is very concentrated.**

   (a) What effect does this have upon the body?
   (b) How does this effect come about?

ANSWERS TO TEST ONE

1. (a) Renal arteries.   (d) Ureter.
   (b) Kidneys.   (e) Bladder.
   (c) Renal veins.   (f) Urethra.

2. (c) Sodium.

3.

	In the cortex of the kidney	In the medulla of the kidney
Bowman's capsule.	(√)	( )
The distal convoluted tubule.	(√)	( )
The proximal convoluted tubule.	(√)	( )
The loop of Henle.	( )	(√)
The glomerulus.	(√)	( )

4. (c) 1300 ml.

5. (b) 120 ml.

6. (b) Potassium ions, and (d) Acid ions.

7. (a) The proximal tubule   (iii) reabsorbs 85% of the volume of the glomerular filtrate.
   (b) The loop of Henle   (i) creates high osmotic pressures in the medulla.
   (c) The distal tubule   (ii) secretes ammonium ions into the tubule.

8. (a) Where there is water deprivation concentrated urine affords the body both maximum retention of water and maximum elimination of waste products.

   (b) The body is able to protect itself in this fashion thanks to the nature of the cells lining the loop of Henle; they permit passage of sodium ions, but not of water.

3. The regulation of urinary losses of water and sodium

3.1. Introduction

The regulation of the water content, and therefore the volume of the extracellular fluid is vital, for if this volume varies, the blood pressure and circulation cannot be maintained.

The regulation of the sodium concentration of extracellular fluid is also essential if the body's cells are to function properly.

The kidney controls the losses of water and sodium in the urine, and thus maintains the **volume**, **sodium ion concentration**, and **osmotic pressure** of the extracellular fluid.

As a result of this control the composition of urine varies greatly with diet, water intake, and with losses of water from the body by other routes (e.g. sweating, respiration).

Typical values for the composition of the urine of a normal person are:

Water — about 1.5 litres per day, but varies greatly with intake
Sodium — 0 to 5 g per day, depending on dietary input
Chloride — about 8 g per day
Potassium — about 2 g per day
Sulphate — about 2 g per day
Urea — about 30 g per day
Glucose — none
Protein — none.

The specific gravity of urine varies between 1.005 and 1.035 depending on the solute concentration.

The pH of urine is about 6.0; that is, urine is slightly acid.

The amounts of water and sodium lost in the urine are under the control of two hormones:

(a) antidiuretic hormone (ADH);
(b) aldosterone.

These hormones exert their effects upon the cells of the distal tubules and collecting ducts.

3.2. Antidiuretic hormone (ADH)

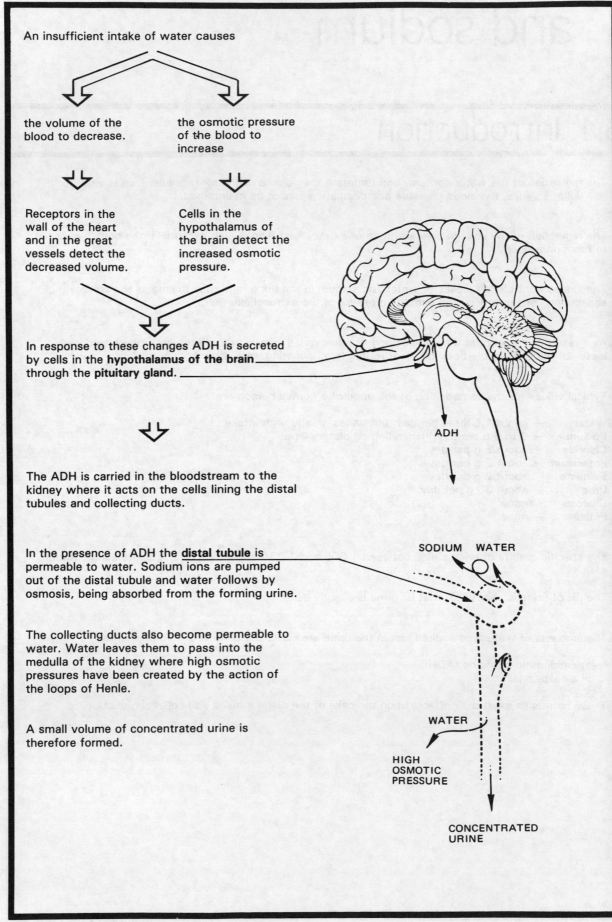

An insufficient intake of water causes

the volume of the blood to decrease.

the osmotic pressure of the blood to increase

Receptors in the wall of the heart and in the great vessels detect the decreased volume.

Cells in the hypothalamus of the brain detect the increased osmotic pressure.

In response to these changes ADH is secreted by cells in the **hypothalamus of the brain** through the **pituitary gland**.

ADH

The ADH is carried in the bloodstream to the kidney where it acts on the cells lining the distal tubules and collecting ducts.

In the presence of ADH the **distal tubule** is permeable to water. Sodium ions are pumped out of the distal tubule and water follows by osmosis, being absorbed from the forming urine.

SODIUM   WATER

The collecting ducts also become permeable to water. Water leaves them to pass into the medulla of the kidney where high osmotic pressures have been created by the action of the loops of Henle.

WATER

A small volume of concentrated urine is therefore formed.

HIGH OSMOTIC PRESSURE

CONCENTRATED URINE

An excessive intake of water causes

the volume of the blood to increase.

the osmotic pressure of the blood to decrease.

Receptors in the wall of the heart and in the great vessels detect the increased volume.

Cells in the hypothalamus of the brain detect the decreased osmotic pressure.

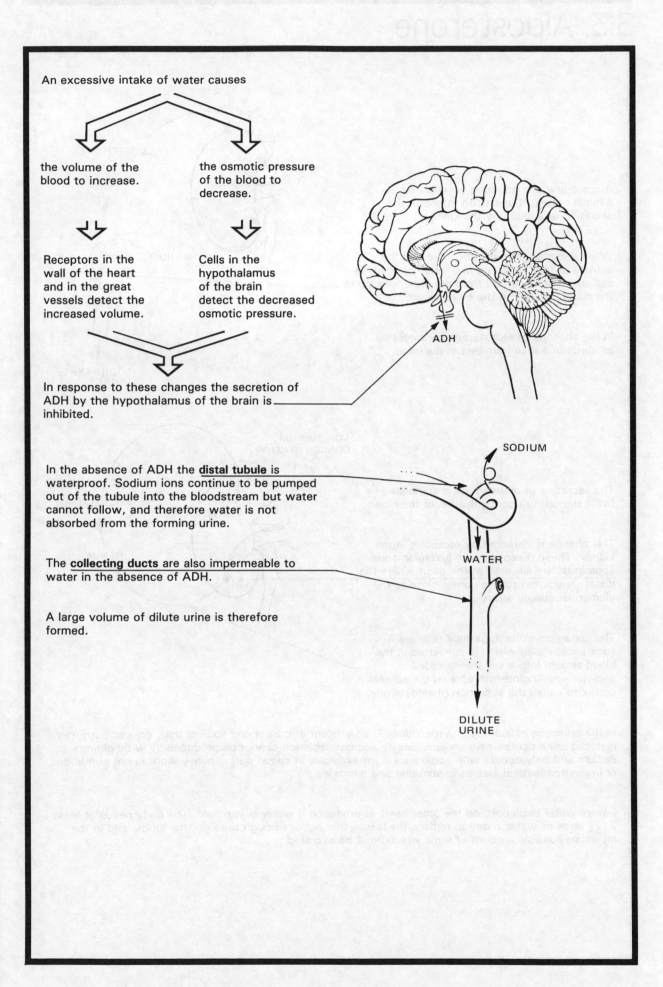

ADH

In response to these changes the secretion of ADH by the hypothalamus of the brain is inhibited.

SODIUM

In the absence of ADH the **distal tubule** is waterproof. Sodium ions continue to be pumped out of the tubule into the bloodstream but water cannot follow, and therefore water is not absorbed from the forming urine.

WATER

The **collecting ducts** are also impermeable to water in the absence of ADH.

A large volume of dilute urine is therefore formed.

DILUTE URINE

3.3. Aldosterone

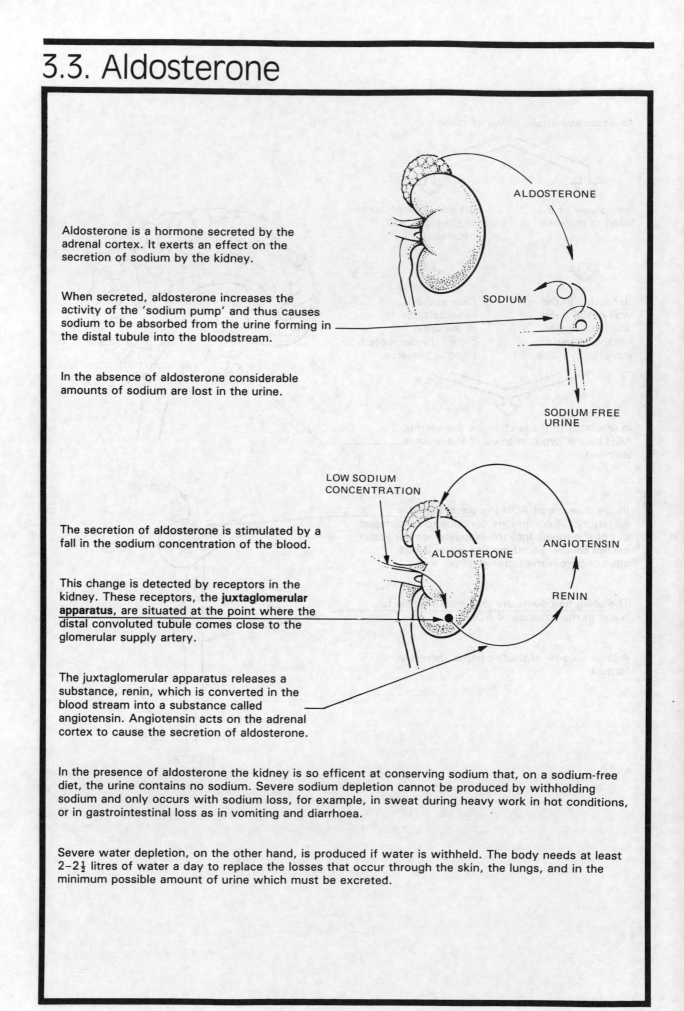

Aldosterone is a hormone secreted by the adrenal cortex. It exerts an effect on the secretion of sodium by the kidney.

ALDOSTERONE

When secreted, aldosterone increases the activity of the 'sodium pump' and thus causes sodium to be absorbed from the urine forming in the distal tubule into the bloodstream.

SODIUM

In the absence of aldosterone considerable amounts of sodium are lost in the urine.

SODIUM FREE URINE

LOW SODIUM CONCENTRATION

The secretion of aldosterone is stimulated by a fall in the sodium concentration of the blood.

ANGIOTENSIN

ALDOSTERONE

This change is detected by receptors in the kidney. These receptors, the **juxtaglomerular apparatus**, are situated at the point where the distal convoluted tubule comes close to the glomerular supply artery.

RENIN

The juxtaglomerular apparatus releases a substance, renin, which is converted in the blood stream into a substance called angiotensin. Angiotensin acts on the adrenal cortex to cause the secretion of aldosterone.

In the presence of aldosterone the kidney is so efficent at conserving sodium that, on a sodium-free diet, the urine contains no sodium. Severe sodium depletion cannot be produced by withholding sodium and only occurs with sodium loss, for example, in sweat during heavy work in hot conditions, or in gastrointestinal loss as in vomiting and diarrhoea.

Severe water depletion, on the other hand, is produced if water is withheld. The body needs at least 2–2½ litres of water a day to replace the losses that occur through the skin, the lungs, and in the minimum possible amount of urine which must be excreted.

TEST TWO

1. **If collecting ducts are made permeable to water by ADH, arrange the following processes in order of their occurrence.**

   (a) Excessive intake of water causes an increase in blood volume and a decrease in the osmotic pressure of the blood.

   (b) Increase in blood pressure is detected by receptors in the heart wall and great vessels, while the decrease in osmotic pressure is detected by cells in the hypothalamus.

   (c) ADH is secreted by cells in the hypothalamus through the pituitary gland.

   (d) ADH is carried via the bloodstream to the kidney.

   (e) ADH causes the distal tubule to become permeable to water so that sodium ions are pumped out of the distal tubules and water follows by passive diffusion (osmosis).

   (f) A large volume of dilute urine is formed.

2. **Why is it that reduced intake of water causes the body to suffer water depletion, while reduced intake of sodium does not lead to sodium depletion?**

ANSWERS TO TEST TWO

1. (e) ADH causes the distal tubule to become permeable to water so that sodium ions are pumped out of the distal tubules and water follows by passive diffusion (osmosis).
   (c) ADH is secreted by cells in the hypothalamus through the pituitary gland.
   (f) A large volume of dilute urine is formed.
   (d) ADH is carried via the bloodstream to the kidney.
   (a) Excessive intake of water causes an increase in blood volume and a decrease in the osmotic pressure of the blood.
   (b) Increase in blood pressure is detected by receptors in the heart wall and great vessels, while the decrease in osmotic pressure is detected by cells in the hypothalamus.

2. Aldosterone causes the kidney to conserve sodium; the body may then suffer sodium depletion through sweating or gastrointestinal loss, but not through dietary sodium deprivation.

4. The ureters and bladder

4.1. Anatomy

Urine in the **collecting ducts** from the nephrons is emptied at the tip of each **renal papilla**

into the **pelvis of the kidney**,

which drains into the **ureter**. The ureter is a tube about 25 cm long and 5 mm wide.

The walls of the ureter contain smooth muscle which contracts and relaxes to produce peristaltic waves which push the urine towards the bladder.

Urine enters the **bladder** in spurts every 10–15 seconds.

The pelvis of the kidney, the ureter, and the bladder are lined with **transitional epithelium**.

This lining is water resistant (or urine would be reabsorbed), and can stretch greatly without necessarily causing damage.

The **ureters** enter the bladder wall obliquely, so that as the bladder fills, urine is not forced back towards the kidney.

As the bladder fills (up to about 500 ml) the smooth muscle in its wall, the **detrusor**, relaxes. The sensitive area at the base of the bladder, the **trigone**, does not stretch.

The smooth muscle of the **internal sphincter** is usually contracted, maintaining continence.

The urine is emptied from the bladder through the **urethra**.

In males the urethra is surrounded by the **prostate gland**.

The **external sphincter** of the urethra is made of striated muscle. It is normally used only occasionally, for example to interrupt the stream of urine voluntarily, but it can take over if the internal sphincter is damaged (for example, after prostate operations).

4.2. The nerve supply of the bladder

The bladder is supplied by autonomic and somatic nerves:

Autonomic nerves

The parasympathetic nerves of the bladder are the **pelvic nerves, (nervi erigentes),** and are the main motor nerves to the detrusor and internal sphincter. They originate in the sacral segments of the spinal cord, in the **vesical centre.**

The **nerves of sensation** from stretch receptors in the bladder wall also run in the pelvic nerves.

Sympathetic nerves from the hypogastric plexus supply muscle around the trigone, and blood vessels in the bladder. They have no important effect on micturition.

Somatic nerves

The somatic nerves are the **pudendal nerves.**

These nerves also originate in the spinal cord, but they take a different route to supply the external sphincter.

The **nerves of sensation** from the urethra and trigone run with the pudendal nerves.

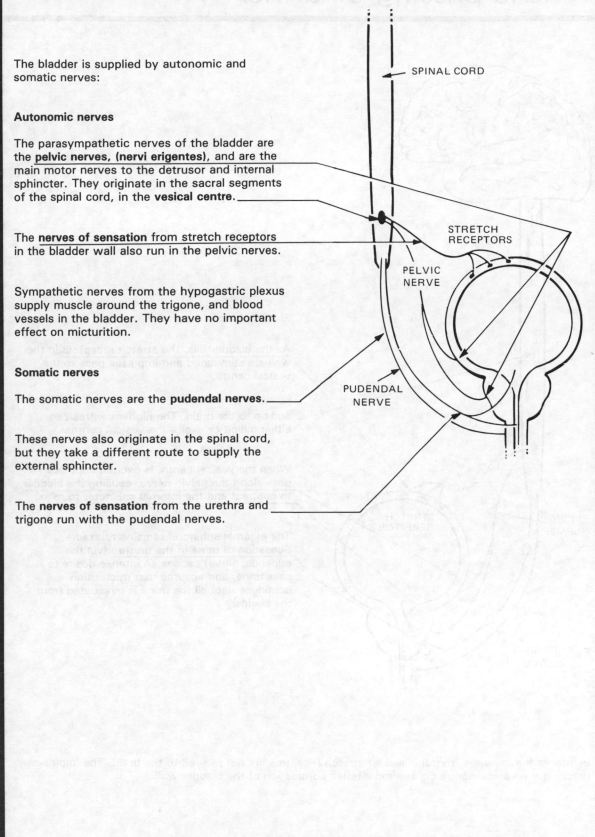

SPINAL CORD

STRETCH RECEPTORS

PELVIC NERVE

PUDENDAL NERVE

4.3. Nervous control of micturition (the passing of urine)

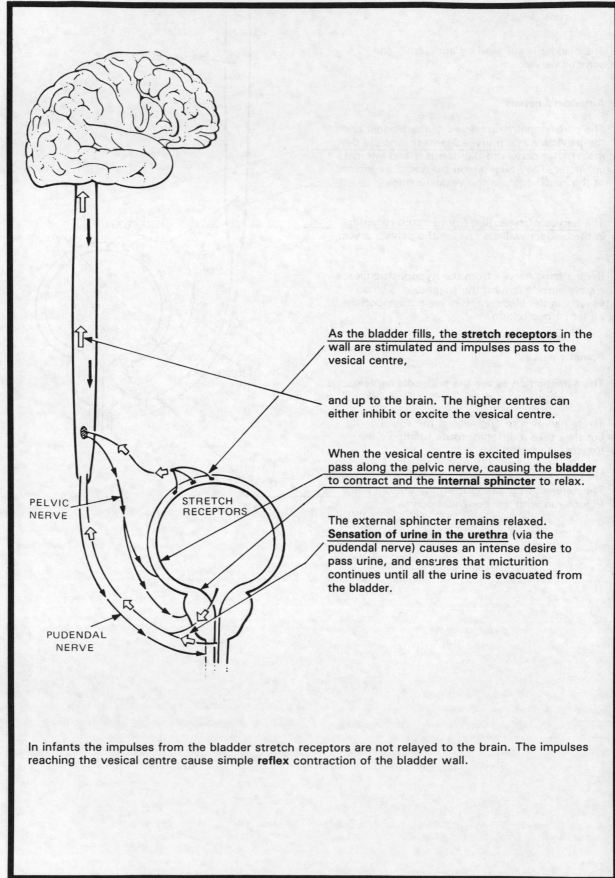

As the bladder fills, the **stretch receptors** in the wall are stimulated and impulses pass to the vesical centre,

and up to the brain. The higher centres can either inhibit or excite the vesical centre.

When the vesical centre is excited impulses pass along the pelvic nerve, causing the **bladder** to contract and the **internal sphincter** to relax.

The external sphincter remains relaxed. **Sensation of urine in the urethra** (via the pudendal nerve) causes an intense desire to pass urine, and ensures that micturition continues until all the urine is evacuated from the bladder.

PELVIC NERVE

STRETCH RECEPTORS

PUDENDAL NERVE

In infants the impulses from the bladder stretch receptors are not relayed to the brain. The impulses reaching the vesical centre cause simple **reflex** contraction of the bladder wall.

TEST THREE

1. **Why is it that when the bladder is very full, urine is not forced back up the ureters to the kidneys?**

   (a) Because there are valves in the ureters.
   (b) Because the ureters enter the bladder wall at such an oblique angle.
   (c) Because of the waves of peristalsis in the ureters.

2. **The stretch receptors in the bladder wall send impulses to the vesical centre:**

   (a) via the pelvic nerve;   (b) via the pudendal nerve.

3. **Characterise the muscle groups that cause micturition.**

	Muscle of bladder wall	Internal sphincter	External sphincter
Smooth muscle.	( )	( )	( )
Striated muscle.	( )	( )	( )
Under voluntary control.	( )	( )	( )
Stimulated by impulses from pelvic nerve.	( )	( )	( )
Stimulated by impulses from pudendal nerve.	( )	( )	( )

ANSWERS TO TEST THREE

1. (a) Because the ureters enter the bladder wall at such an oblique angle.
   (b) Because of the waves of peristalsis in the ureters.

2. (a) Via the pelvic nerve.

3.

	Muscle of bladder wall	Internal sphincter	External sphincter
Smooth muscle.	(√)	(√)	( )
Striated muscle.	( )	( )	(√)
Under voluntary control.	( )	( )	(√)
Stimulated by impulses from pelvic nerve.	(√)	(√)	( )
Stimulated by impulses from pudendal nerve.	( )	( )	(√)

POST TEST

1. (a) How are the composition and volume of extracellular fluid controlled?
   (b) Why is it important that they should be controlled?

   _____

   _____

2. Indicate which of the names in the list below refer to the parts of the kidney labelled on the diagram alongside, by placing the appropriate letters in the brackets.

   1. The arteriole.   ( )
   2. Bowman's capsule.   ( )
   3. The collecting duct.   ( )
   4. The distal convoluted tubule.   ( )
   5. The glomerulus.   ( )
   6. The loop of Henle.   ( )
   7. The proximal convoluted tubule.   ( )

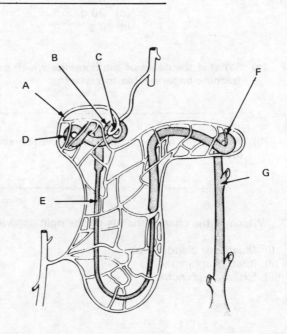

3. Which of the descriptions listed on the right apply to the parts of the kidney listed on the left?

   (i)   The glomerulus.   (a) Passes through the medulla.
   (ii) The proximal tubule. (b) Deals with about 15% of the filtered fluid.
   (iii) The distal tubule.   (c) Depends on an adequate blood pressure to function properly.
   (iv) The collecting duct. (d) Absorbs most of the glucose.

4. Which of the following filter through the glomerulus?

   (a) Sodium ions.   (b) Red blood cells.   (c) Glucose.   (d) Creatinine.   (e) Proteins.

5. Approximately how much of the following substances is excreted each day in the urine of a normal person?

(i) Potassium.   (a) 5 g.
(ii) Protein.   (b) 2 g.
(iii) Urea   (c) None.
                     (d) 30 g.
                     (e) 8 g.

6. (a) What is the name of the hormone which causes the distal tubules and collecting ducts to become impermeable to water?

_____

(b) What is the name of the hormone which stimulates the 'sodium pump' activity in the distal tubule?

_____

7. Which of the characteristics on the right apply to the muscles of the bladder listed on the left?

(i) Muscle of bladder wall.   (a) Smooth muscle.
(ii) Internal sphincter.   (b) Striated muscle.
(iii) External sphincter.   (c) Stimulated via the autonomic nervous system.
                           (d) Stimulated via the somatic nerves.
                           (e) Supplied by the pudendal.
                           (f) Supplied by the pelvic nerve.

8. Which of the descriptions on the right apply to the parts of the urinary system listed on the left?

(i) Ureter.   (a) Encircled by the prostate in males.
(ii) Detrusor.   (b) Forms the base of the bladder.
(iii) Trigone.   (c) Can contract voluntarily.
(iv) External sphincter.   (d) Contracts when stimulated by the pelvic nerve.
(v) Urethra.   (e) Produces peristaltic waves.

ANSWERS TO POST TEST

1. (a) The kidneys control the composition and volume of extracellular fluid.
    (b) The cells require a closely regulated environment for functioning properly.

2. 1. The arteriole.   (A)
    2. Bowman's capsule.   (B)
    3. The collecting duct.   (G)
    4. The distal convoluted tubule.   (F)
    5. The glomerulus.   (C)
    6. The loop of Henle.   (E)
    7. The proximal convoluted tubule.   (D)

3.   (i)   The glomerulus   (c) depends on an adequate blood pressure to function
                                       properly.
      (ii) The proximal tubule   (d) absorbs most of the glucose.
      (iii) The distal tubule   (b) deals with about 15% of the filtered fluid.
      (iv) The collecting duct   (a) passes through the medulla.

4. (a) Sodium ions.   (c) Glucose.   (d) Creatinine.

ANSWERS TO POST TEST

5. (i) (b) 2 g.
        (ii) (c) None.
        (iii) (d) 30 g.

6. (a) Antidiuretic hormone (ADH).
        (b) Aldosterone.

7. (i) The muscle of bladder wall is (a) smooth muscle (c) stimulated via the autonomic nervous system (f) supplied by the pelvic nerve.

        (ii) The internal sphincter is (a) smooth muscle (c) stimulated via the autonomic nervous system (f) supplied by the pelvic nerve.

        (iii) The external sphincter is (b) striated muscle (d) stimulated via the somatic nerves (e) supplied by the pudendal nerve.

8. (i) The ureter.   (e) produces peristaltic waves.
        (ii) The detrusor   (d) contracts when stimulated by the pelvic nerve.
        (iii) The trigone   (b) forms the base of the bladder.
        (iv) The external sphincter   (c) can contract voluntarily.
        (v) The urethra   (a) is encircled by the prostate in males.

Contents

The Digestive System

1. Introduction

The body needs a system to convert food into a form usable by its cells. The body's cells can only utilise a very narrow range of substances. It is the function of the digestive system to reduce foods to these substances and allow for their absorption into the body.

Food consists of:

proteins — long and complex chains of simple nitrogen-containing compounds called amino acids. There are about 20 amino acids which are essential to life;

PROTEIN

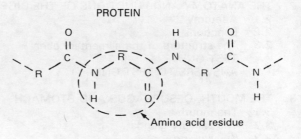

Amino acid residue

fats — compounds formed from simple organic acids (fatty acids) and glycerol. The most common fats are *triglycerides*;

TRIGLYCERIDE FAT

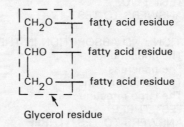

Glycerol residue

carbohydrates
— *monosaccharides* simple hexose sugars such as glucose, fructose, galactose; *polysaccharides* combinations of two or more sugar molecules such as starch, which consists of long chains of glucose molecules;

STARCH

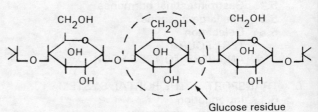

Glucose residue

water — the body needs sufficient water to ensure an adequate urine flow, and to replace losses due to evaporation from the skin and in the breath. About $2\frac{1}{2}$ litres a day is usually sufficient in temperate conditions, but actual requirements vary widely.

salts — simple substances present in solution as electrically charged particles called *ions*. A salt, or electrolyte, solution is always electrically neutral, but consists of a mixture of positively charged metal ions *(cations)* and negatively charged non-metallic ions *(anions)*. Solutions of salts form the basic body fluids in which the more complex constituents of the body exist. The maintenance of the body fluids at the correct concentrations is the task of the kidneys. The digestive system must absorb sufficient salts to enable losses to be made good.

Salts are usually measured, by their equivalent chemical activity, as milliequivalents (mEq). The most important ions in the body are the following.

	MAJOR LOCATION IN THE BODY	TOTAL BODY CONTENT (ADULT)	CONC. IN BLOOD PLASMA	AVERAGE DAILY INTAKE	SOURCE IN DIET
SODIUM Na^+	The body fluids *outside* the cells (extracellular fluid) including blood plasma.	200–250 g (as common salt NaCl) 3500 mEq.	140 mEq/l	Varies widely 10–20 g NaCl	Almost all foods, and as common salt itself.
POTASSIUM K^+	The body fluid *within* the cells (intracellular fluid).	180–230 g (as KCl) 3200 mEq	4 mEq/l	Varies widely	Almost all food especially *meat* and *fruit*.
CALCIUM Ca^{++}	99% in bone as crystals of calcium salts.	1200 g	10 mEq/l	500–1000 mg Vitamin D needed for absorption. More needed by children than adults.	Milk and milk products, cereals. Hard water.
IRON Fe^{++}	In haemoglobin, the red oxygen — carrying pigment of blood, and in other body enzymes.	5 g	100 μg /100 ml	10–20 mg (More needed by menstruating women, male requirement very low).	Meat, wholemeal cereal, red wine.
IODINE $I-$	The thyroid gland and in blood as thyroid hormone.		5–10 μg /100 ml (as hormone).	150 μg	Fish — and as additive in table salt.

vitamins — substances essential to cellular function which the body itself cannot manufacture, and which are only needed in small amounts. The most important vitamins are the following:

FAT SOLUBLE VITAMINS	SOURCE	FUNCTION	DAILY REQUIREMENT	EFFECT OF DEFICIENCY
Vitamin A (Retinol)	Milk, eggs, fish liver oils.	Needed by all active body cells, including skin and eyes.	800 μg	Dry skin. Retinal disorder with night blindness, corneal ulceration with ultimate destruction of eye.
Vitamin D	Fish, milk products (Made by skin in bright sunlight).	Absorption, metabolism of calcium, growth and maintenance of bone.	10 μg	Rickets in children. Osteomalacia (bone softening) in adults. Tetany (muscle spasms) if severe.
Vitamin K	Green vegetables.	Acid blood clotting.		Rare (but found in *jaundice* when absorption is poor). Generalised bleeding.

WATER SOLUBLE VITAMINS	SOURCE	FUNCTION	DAILY REQUIREMENT	EFFECT OF DEFICIENCY
Vitamin B1 Thiamine	Liver, wheatgerm, nuts and yeast	Carbohydrate metabolism in all cells.	1−2 mg	Beri-beri (with heart failure), nerve disorders and confusional states.
Niacin	Fish, liver wheatgerm.	Energy metabolism in all cells.	12 mg	Pellagra (skin rash, swollen tongue, diarrhoea and mental confusion).
Riboflavin and Pyridoxine	Liver, kidney, milk and milk products, eggs.	Energy and protein metabolism.	5 mg 2 mg	Rare (usually part of multiple vitamin deficiency) affects skin and nerve function.
Vitamin B12	Meat, inc. liver.	DNA synthesis.	2 μg	Pernicious anaemia, spinal cord disease.
Folate	Green vegetables, kidney, liver, fish.	DNA synthesis		Anaemia, bowel disorders.
Vitamin C (Ascorbic acid)	Citrus fruits, other fruits, green vegetables.	Formation of collagen and intercellular substance by connective tissue.	10 mg minimum to prevent scurvy. 20−30 mg recommended.	Scurvy, with swollen gums, ready bruising, weakness and failure of normal healing.

In the digestive system

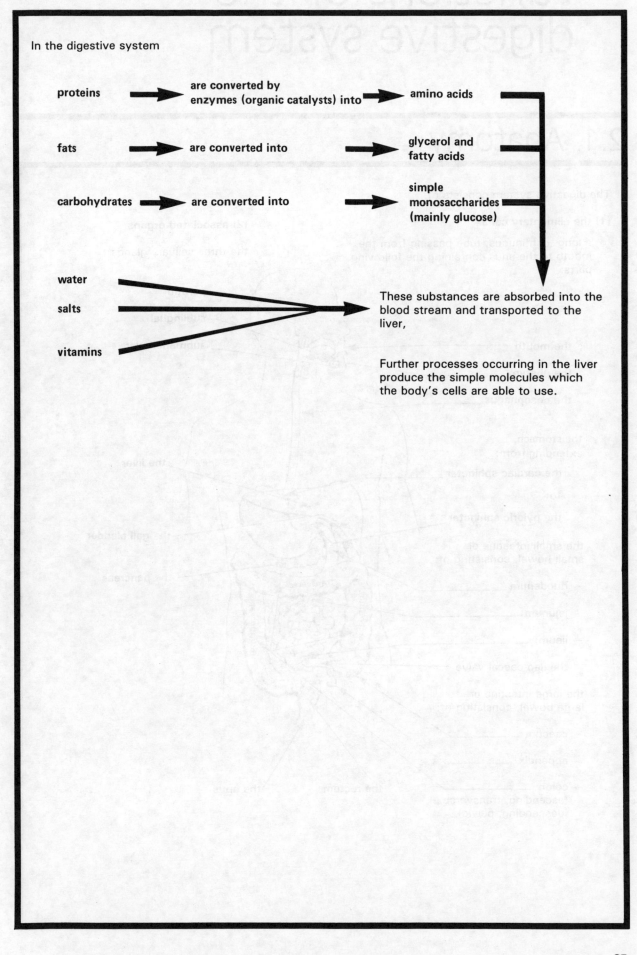

proteins → are converted by enzymes (organic catalysts) into → amino acids

fats → are converted into → glycerol and fatty acids

carbohydrates → are converted into → simple monosaccharides (mainly glucose)

water
salts
vitamins

These substances are absorbed into the blood stream and transported to the liver,

Further processes occurring in the liver produce the simple molecules which the body's cells are able to use.

2. The anatomy and functions of the digestive system

2.1. Anatomy

The digestive system consists of:

(1) the alimentary canal

a long, continuous, tube passing from the mouth to the anus containing the following parts:

the **mouth**

the **pharynx**

the **oesophagus**

the **stomach**, extending from:

the **cardiac sphincter**

to:

the **pyloric sphincter**

the small intestine or small bowel, consisting of:

— **duodenum**

— **jejunum**

— **ileum;**

the **ileo-caecal valve**

the large intestine or large bowel, consisting of:

— **caecum**

— **appendix**

— **colon**
(ascending, transverse, descending, pelvic);

(2) associated organs:

the three salivary glands

parotid

sublingual

submandibular

the liver

the gall bladder

the pancreas

transverse

ascending

descending

the rectum   **the anus**

2.2. Functions

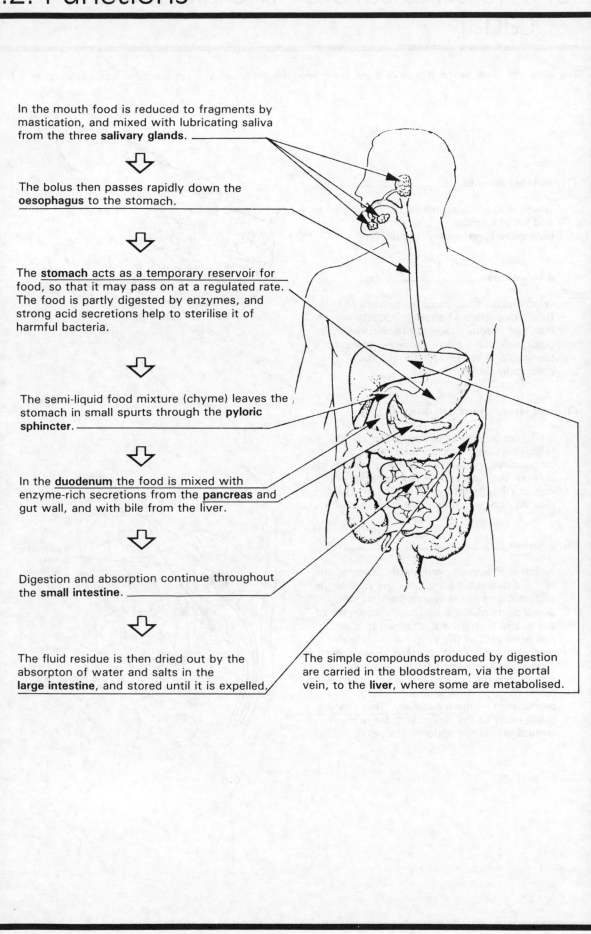

In the mouth food is reduced to fragments by mastication, and mixed with lubricating saliva from the three **salivary glands**.

⇩

The bolus then passes rapidly down the **oesophagus** to the stomach.

⇩

The **stomach** acts as a temporary reservoir for food, so that it may pass on at a regulated rate. The food is partly digested by enzymes, and strong acid secretions help to sterilise it of harmful bacteria.

⇩

The semi-liquid food mixture (chyme) leaves the stomach in small spurts through the **pyloric sphincter**.

⇩

In the **duodenum** the food is mixed with enzyme-rich secretions from the **pancreas** and gut wall, and with bile from the liver.

⇩

Digestion and absorption continue throughout the **small intestine**.

⇩

The fluid residue is then dried out by the absorpton of water and salts in the **large intestine**, and stored until it is expelled.

The simple compounds produced by digestion are carried in the bloodstream, via the portal vein, to the **liver**, where some are metabolised.

2.3. The structure of the alimentary canal

The different parts of the digestive tube, from oesophagus to anus, have a similar basic structure:

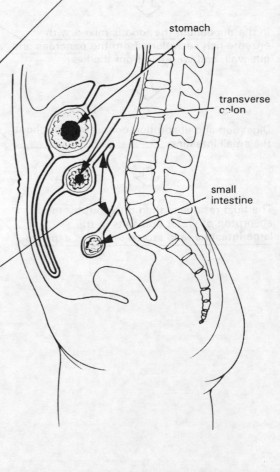

(1) an inner mucosa

which is specialised for *secretion* by glands, and for *absorption*. It also acts as a protective layer against bacteria;

(2) a submucosa

which forms the strong framework of the tube. It consists of a tough, closely woven mesh of fibrous tissue containing the main blood vessels, a nerve network (Meissner's plexus), and in the duodenum, alkali-producing glands;

(3) two layers of smooth muscle

a circular muscle (a close-packed spiral) which can constrict the tube, and a longitudinal muscle (a long spiral) which can shorten the tube. There is a network of nerves (the myenteric plexus or Auerbach's plexus) between these two muscle layers;

(4) a serosa

which is a layer of peritoneum covering the surface of the tube and also the wall of the abdominal cavity in which the tube lies. Some parts of the tube (duodenum, part of the colon, rectum) are attached to the posterior wall of the cavity and are only partly covered. Other parts (stomach, part of the small intestine, transverse colon) lie free within the cavity, and are supplied with blood through a thick double fold of peritoneum — **the mesentery**. This enables these parts of the digestive tube to move around within the abdominal cavity.

stomach

transverse colon

small intestine

TEST ONE

1. **Which of the following compounds can be absorbed directly in the digestive tract, without further digestion?**

   (a) Amino acids.   (b) Fats.   (c) Starches.   (d) Enzymes.   (e) Glucose.   (f) Vitamins.

2. **Complete the following:**

   (a) Water, salts, and ____ are processed by the ____ into substances usable by the body's cells.
   (b) ____ are converted into ____ monosaccharides.
   (c) Fats are converted into ____ and ____ acids.
   (d) ____ are converted into ____ acids.

3. **At which of the following parts of the alimentary canal does a significant amount of absorption occur?**

   (a) The mouth.   (b) The oesophagus.   (c) The stomach.   (d) The small intestine.   (e) The large intestine.

4. **Which of the descriptions on the right apply to the parts of the digestive system listed on the left?**

   (a) The oesophagus.   (i) The exit from the stomach.
   (b) The stomach.   (ii) Carries food to the stomach.
   (c) The pylorus.   (iii) Produces an enzyme-rich secretion.
   (d) The pancreas.   (iv) Serves as a temporary food reservoir.
   (e) The large bowel.   (v) Absorbs water and salts.

5. **Which of the descriptions on the right apply to the parts of the digestive tube listed on the left?**

   (a) The mucosa.   (i) Enables the bowel to move.
   (b) The sub-mucosa.   (ii) Specialised for absorption and secretion.
   (c) The smooth muscle layers. (iii) Covers the surface of the bowel.
   (d) The peritoneum.   (iv) A double layer of serosa carrying blood vessels.
   (e) The mesentery.   (v) The strongest layer in the tube.

ANSWERS TO TEST ONE

1. (a) Amino acids.   (e) Glucose.   (f) Vitamins.

2. (a) Water, salts, and *vitamins* are processed by the *liver* into substances usable by the body's cells.
   (b) *Carbohydrates* are converted into *simple monosaccharides*.
   (c) Fats are converted into *glycerol* and *fatty* acids.
   (d) *Proteins* are converted into *amino* acids.

3. (d) The small intestine.   (e) The large intestine.

4. (a) The oesophagus   (ii) carries food to the stomach.
   (b) The stomach   (iv) serves as a temporary reservoir for food.
   (c) The pylorus   (i) is the exit from the stomach.
   (d) The pancreas   (iii) produces an enzyme-rich secretion.
   (e) The large bowel   (v) absorbs water and salts.

5. (a) The mucosa   (ii) is specialised for absorption and secretion.
   (b) The sub-mucosa   (v) is the strongest layer in the tube.
   (c) The smooth muscle layer (i) enables the bowel to move.
   (d) The peritoneum   (iii) covers the surface of the bowel.
   (e) The mesentery   (iv) is a double layer of serosa carrying blood vessels.

3. The mouth, oesophagus and stomach

3.1. The mouth

The mouth is lined with stratified squamous epithelium to resist wear and tear.

The predominant taste of food is appreciated by receptors on the tongue; the appreciation of more subtle *flavours* involves the sense of smell. The consistency of food is detected by delicate touch receptors on the tongue, which provide a safeguard against the swallowing of harmful substances. Food is held between the teeth by the tongue and cheek muscles.

Food is broken up by the teeth. There are 32 teeth in the second dentition.
In each half-jaw there are:

two sharp **incisors** for cutting;

one spike-like **canine** for tearing;

two **premolars**;

and three **molars** for grinding.

3.2. The salivary glands

The sight, smell and taste of food provokes the secretion of saliva by a nerve reflex. Saliva lubricates the food and enables it to be converted into a soft mass, or *bolus*. Some food dissolves in the saliva and the dissolved substances can then more readily stimulate the taste receptors. In addition to these functions, saliva contains an enzyme, *ptyalin*, which starts the breakdown of starch into simple sugars.

Saliva is secreted by three major glands:

the **paratoid gland** which produces a watery saliva;

the **sublingual gland**, and
the **submandibular gland**, which produce a mixed watery and mucous saliva;

and by many small salivary glands scattered throughout the cheeks and palate.

About 1–1½ litres of saliva are produced every day.

3.3. Swallowing

Swallowing begins as a voluntary act which merges smoothly into an involuntary reflex. It takes place in three stages.

1. *Buccal stage*

   Food is gathered on the upper surface of the **tongue** as a moist **bolus**.

   The tongue is then pressed against the **hard palate** forcing the bolus backwards.

   The **soft palate** is raised to prevent the food entering the nose, and the bolus is forced into the **pharynx**.

2. *Pharyngeal stage*

   The **larynx** is pulled up under the base of the tongue, the laryngeal inlet constricts, and the **epiglottis** folds over the larynx to prevent food entering the trachea.

   The **cricopharyngeal** sphincter between the pharynx and the oesophagus is usually closed to prevent air being drawn into the oesophagus during respiration, but this relaxes briefly as the bolus reaches it.

   The muscles of the pharynx then squeeze the bolus into the upper **oesophagus**.

3. *Oesophageal stage*

   Peristaltic waves carry the food bolus down to the stomach.

3.4. The oesophagus

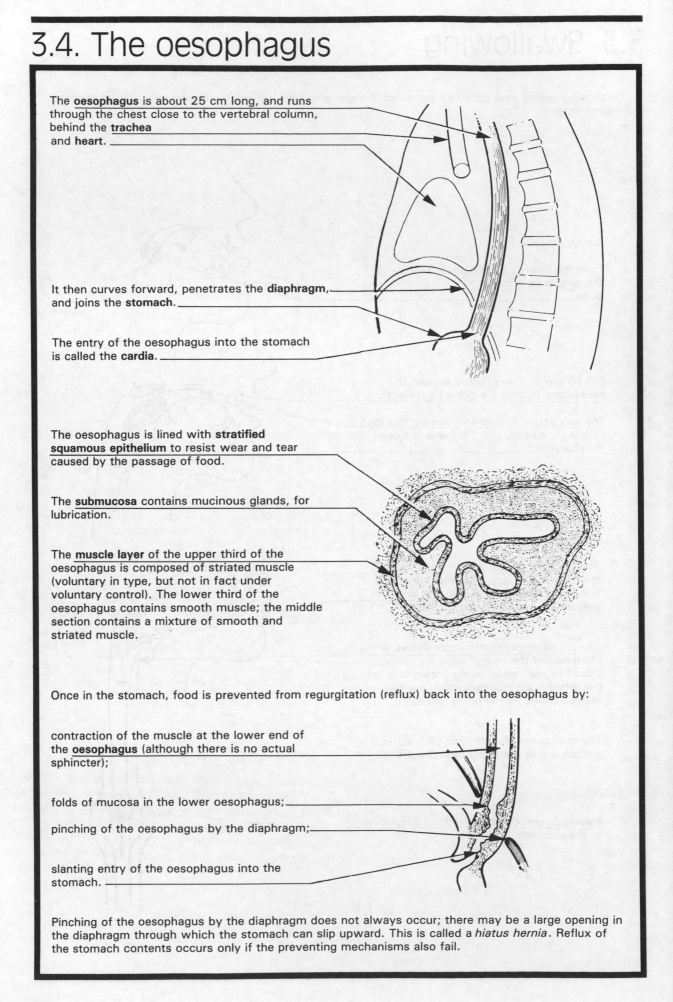

The **oesophagus** is about 25 cm long, and runs through the chest close to the vertebral column, behind the **trachea** and **heart**.

It then curves forward, penetrates the **diaphragm**, and joins the **stomach**.

The entry of the oesophagus into the stomach is called the **cardia**.

The oesophagus is lined with **stratified squamous epithelium** to resist wear and tear caused by the passage of food.

The **submucosa** contains mucinous glands, for lubrication.

The **muscle layer** of the upper third of the oesophagus is composed of striated muscle (voluntary in type, but not in fact under voluntary control). The lower third of the oesophagus contains smooth muscle; the middle section contains a mixture of smooth and striated muscle.

Once in the stomach, food is prevented from regurgitation (reflux) back into the oesophagus by:

contraction of the muscle at the lower end of the **oesophagus** (although there is no actual sphincter);

folds of mucosa in the lower oesophagus;

pinching of the oesophagus by the diaphragm;

slanting entry of the oesophagus into the stomach.

Pinching of the oesophagus by the diaphragm does not always occur; there may be a large opening in the diaphragm through which the stomach can slip upward. This is called a *hiatus hernia*. Reflux of the stomach contents occurs only if the preventing mechanisms also fail.

3.5. The stomach

The stomach is a muscular sac. The different parts of the stomach have specific functions.

The **fundus** is relatively thin-walled, has few glands and serves as a reservoir.

The **body** is muscular and stores and mixes the food.

The **body** contains many gastric glands and is the only site of acid secretion by gastric glands.

The **pyloric sphincter** guards the exit from the stomach — the pylorus.

The **pyloric antrum** is composed of thick muscle and acts as a pump to transport food to the small bowel at a controlled rate, as well as mixing it further.

3.6. Digestion in the stomach

Absorption

Absorption in the stomach is very limited, but alcohol and glucose are well absorbed.

Secretion

In the stomach food is converted by various secretions from gastric glands into a milky fluid called **chyme**, which is suitable for passage through the small intestine.

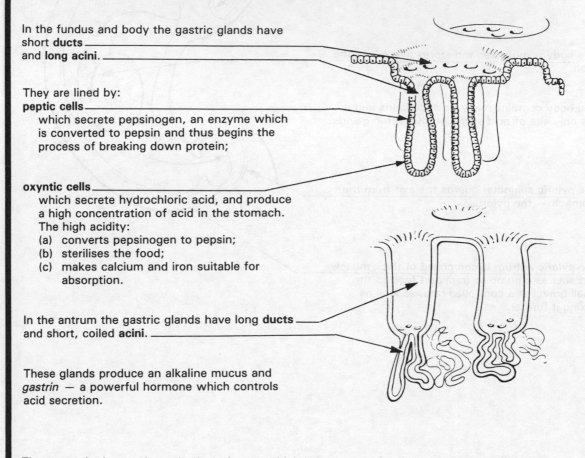

In the fundus and body the gastric glands have short **ducts** and **long acini**.

They are lined by:
peptic cells
which secrete pepsinogen, an enzyme which is converted to pepsin and thus begins the process of breaking down protein;

oxyntic cells
which secrete hydrochloric acid, and produce a high concentration of acid in the stomach.
The high acidity:
(a) converts pepsinogen to pepsin;
(b) sterilises the food;
(c) makes calcium and iron suitable for absorption.

In the antrum the gastric glands have long **ducts** and short, coiled **acini**.

These glands produce an alkaline mucus and *gastrin* — a powerful hormone which controls acid secretion.

The stomach also produces *intrinsic factor*, which is necessary for the absorption of vitamin B_{12}.

Mucous cells throughout the stomach produce mucus which protects the stomach wall from its own digestive juices.

TEST TWO

1. **What are the functions of the different kinds of teeth?**

		Tearing	Cutting	Grinding
(a)	Incisors.	( )	( )	( )
(b)	Canines.	( )	( )	( )
(c)	Premolars.	( )	( )	( )
(d)	Molars.	( )	( )	( )

2. (a) **What are the stages of swallowing?**
   (b) **Relate the following organs to the appropriate stages.**

   Pharynx.
   Tongue.
   Larynx.
   Soft palate.
   Epiglottis.
   Upper oesophagus.
   Cricopharangeal sphincter.
   Hard palate.

3. **List three factors which prevent regurgitation of food from the stomach up the oesophagus.**

   (i) _____

   (ii) _____

   (iii) _____

4. **Which of the descriptions on the right apply to the parts of the stomach listed on the left?**

   (a) The cardia.   (i) Thin-walled.
   (b) The fundus.   (ii) Contains a muscular sphincter.
   (c) The body.   (iii) The entrance to the stomach.
   (d) The antrum.   (iv) The site of gastrin production.
   (e) The pylorus.   (v) The main site of acid secretion.

5. **What are the main functions of the hydrochloric acid which is secreted by the gastric glands?**

   _____

   _____

   _____

ANSWERS TO TEST TWO

1. **What are the functions of the different kinds of teeth?**

		Tearing	Cutting	Grinding
(a)	Incisors	( )	(√)	( )
(b)	Canines	(√)	( )	( )
(c)	Premolars	( )	( )	(√)
(d)	Molars	( )	( )	(√)

2. (a) The stages of swallowing are the buccal, the pharyngeal and the oesophageal.

   (b) **Buccal stage**

   Tongue.
   Hard palate.
   Soft palate.
   Pharynx.

   Pharyngeal stage

   Larynx.
   Epiglottis.
   Cricopharyngeal sphincter.
   Upper oesophagus.

3. Three of the following:

   Contraction of muscle at the lower end of the oesophagus.
   Folds of mucosa at the lower end of the oesophagus.
   Pinching of the oesophagus by the diaphragm.
   Slanting entry of oesophagus into stomach.

4. (a) The cardia   (iii) is the entrance to the stomach.
   (b) The fundus   (i) is thin-walled.
   (c) The body   (v) is the main site of acid secretion.
   (d) The antrum   (iv) is the site of gastrin production.
   (e) The pylorus   (ii) contains a muscular sphincter.

5. **Hydrochloric acid**

   (i) converts pepsinogen to pepsin;
   (ii) sterilises food;
   (iii) makes calcium and iron suitable for absorption.

4. Digestion and absorption in the intestine

4.1. The duodenum

The **duodenum** is the first 25 cm of the small bowel, and is wrapped around the head of the **pancreas**.

Chyme entering the duodenum through the pylorus is mixed with:

(1) secretions of the duodenal wall;
(2) bile;
(3) pancreatic juice.

Duodenal secretions from the mucosal glands and from the submucosal *Brunner's glands*, contain bicarbonate and are alkaline, thus helping to neutralise the acid chyme.

Bile (600 ml/day) is secreted by the cells of the liver, and is stored and concentrated (about tenfold) in the **gall bladder**.

The presence of food (especially fat) in the duodenum causes the gall bladder to contract and empty bile down the **cystic duct** and **bile duct** through the **ampulla**.

Bile is made up of the following:

(1) *Bile salts*

Combinations of cholesterol-based acids, such as cholic acid, with the amino acids taurine and glycine, make up $\frac{2}{3}$ of the dry weight of bile. These compounds act as wetting agents, and emulsify fats into very fine droplets, or micelles, aiding their absorption.

Bile salts also activate lipase, a pancreatic enzyme, and stimulate pancreatic secretion.

(2) *Bile pigments*

These consist mainly of bilirubin, a haemoglobin breakdown product from the bone marrow, which is excreted by the liver and colours the stool.

(3) *Cholesterol, lecithin, salts* and *water*.

4.2. The pancreas

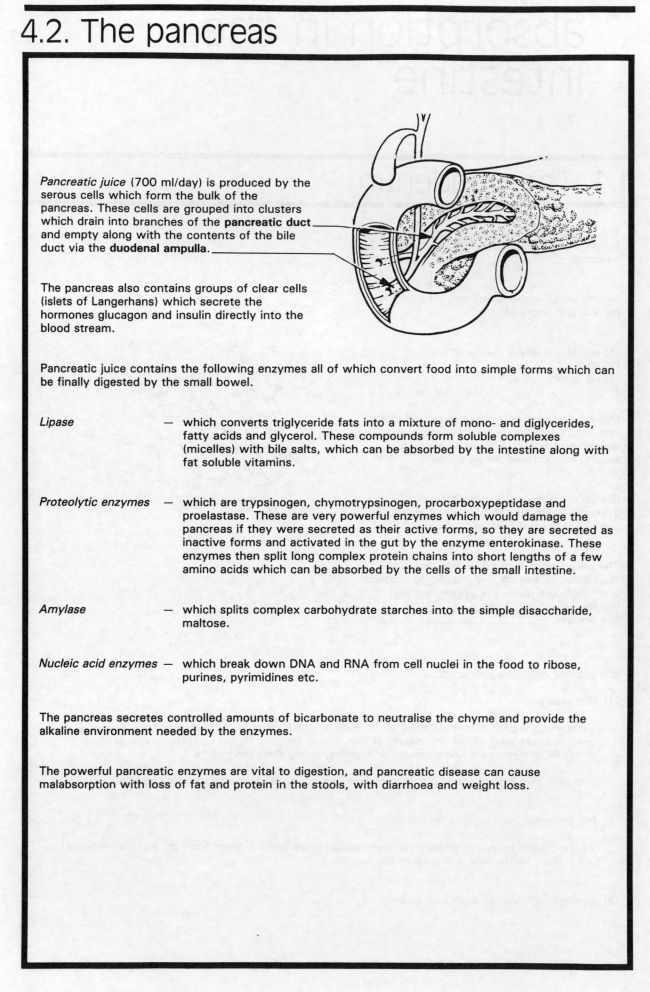

Pancreatic juice (700 ml/day) is produced by the serous cells which form the bulk of the pancreas. These cells are grouped into clusters which drain into branches of the **pancreatic duct** and empty along with the contents of the bile duct via the **duodenal ampulla.**

The pancreas also contains groups of clear cells (islets of Langerhans) which secrete the hormones glucagon and insulin directly into the blood stream.

Pancreatic juice contains the following enzymes all of which convert food into simple forms which can be finally digested by the small bowel.

Lipase — which converts triglyceride fats into a mixture of mono- and diglycerides, fatty acids and glycerol. These compounds form soluble complexes (micelles) with bile salts, which can be absorbed by the intestine along with fat soluble vitamins.

Proteolytic enzymes — which are trypsinogen, chymotrypsinogen, procarboxypeptidase and proelastase. These are very powerful enzymes which would damage the pancreas if they were secreted as their active forms, so they are secreted as inactive forms and activated in the gut by the enzyme enterokinase. These enzymes then split long complex protein chains into short lengths of a few amino acids which can be absorbed by the cells of the small intestine.

Amylase — which splits complex carbohydrate starches into the simple disaccharide, maltose.

Nucleic acid enzymes — which break down DNA and RNA from cell nuclei in the food to ribose, purines, pyrimidines etc.

The pancreas secretes controlled amounts of bicarbonate to neutralise the chyme and provide the alkaline environment needed by the enzymes.

The powerful pancreatic enzymes are vital to digestion, and pancreatic disease can cause malabsorption with loss of fat and protein in the stools, with diarrhoea and weight loss.

4.3. The small intestine

The small intestine is about 4–7 metres long, the length varying with the contractions and relaxations of its muscular wall.

The mucosa is covered with finger-like **villi**, which give the small bowel a large surface area (about 10m^2). There are about 25–40 villi/mm^2, each villus being about 1 mm long.

In the duodenum and jejunum the mucosa is thrown into folds and the villi are long and tightly packed. Towards the ileum the mucosa is less folded, the wall thinner, and the villi shorter and sparser.

The villi are in constant waving motion.

The cells of the walls of the villi constantly migrate upwards, taking 2–3 days to reach the tips of the villi where they are shed. About 100 g of cells are shed daily by the bowel.

The cells and villi do *not* secrete enzymes, apart from enterokinase, but some enzymes are released on the shedding of cells.

Digestion continues in the lining cells themselves, particularly in the *brush border* and the sticky *glycocalx* which covers them. The brush border is formed by micro-villi which increase the exposed surface area even further.

brush border

glycocalx — nucleus

capillaries

the lining epithelium

lacteal

intestinal gland vein

In the cells lining the villi the following occur.

Proteases break down peptides to amino acids which are absorbed through capillaries into the bloodstream.

Lactase, maltase, sucrase
break down disaccharides to monosaccharides (mainly glucose) which are absorbed through capillaries into the bloodstream.

Lipase
acts on fat micelles to form:
   simple fatty acids and glycerol, which are absorbed through capillaries into the bloodstream;

   long chain fatty acids and glycerol, which recombine to form triglyceride fats and pass into the lymphatic *lacteals* as fine droplets (chylomicrons) along with the fat soluble vitamins A and D;

   bile salts, which are reabsorbed in the lower ileum.

Water soluble vitamins
are absorbed directly into the bloodstream.

Iron
is absorbed mainly in the upper duodenum.

Vitamin B$_{12}$
(linked with intrinsic factor) is absorbed in the lower ileum.

All important digestion and absorption takes place in the small bowel. Both the stomach and the large bowel can be entirely removed without serious ill-health. Up to one-third of the small bowel can be removed without serious effects on digestion, and survival is possible with as little as one metre of small bowel intact.

4.4. The large intestine

About 8 litres of water enter the digestive tract each day from the diet and secretions. Most of this is absorbed by the small bowel.

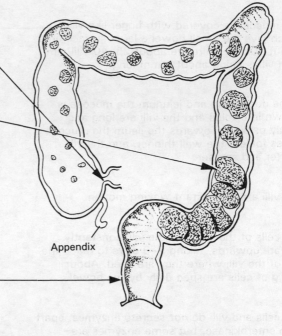

About 1 litre of liquid chyme enters the colon each day, through the **ileo-caecal valve**.

Water and salts, particularly potassium salts, are absorbed throughout the colon. The **faeces** therefore become progressively firmer as they travel towards the rectum.

The mucosa lining the large bowel is rich in mucus-secreting goblet cells. No enzymes are secreted by the mucosa and proteins and fats cannot be absorbed by the large bowel.

However, some drugs, in suppository form, can be absorbed across the **rectal** mucosa.

Appendix

The large bowel contents team with saprophytic and non-pathogenic bacteria. They prevent the growth of dangerous strains of bacteria, and may also be involved in the synthesis of B vitamins. Protection against bacteria is also provided by lymphoid nodules in the sub-mucosa.

Composition of faeces

Normally only one quarter of the faeces is solid matter — the remainder is water.

The solid matter consists of:

dead bacteria	(30%)
indigestible matter, e.g. cellulose	(30%)
inorganic matter, e.g. calcium salts	(10–20%)
shed cells	(50–100 g/day)
leucocytes	
bile pigments.	

Faeces form in almost normal amounts during starvation.

TEST THREE

1. **Label the parts indicated on the diagram below:**

    A

    B

    C

    D

    E

2. (a) **Where is bile produced?** _____

   (b) **Where is bile stored?** _____

3. **What are the three main functions of bile salts?**

   _____

   _____

   _____

4. **Relate the following to the duodenum, the pancreas or the small intestine.**

   Lipase.   The gall bladder.   Maltase.
   Proteases.   Nucleic acid enzymes.   The cystic duct.
   Bile.   Sucrase.   The ampulla.
   Bicarbonate.   Proteolytic enzymes.
                            Amylase.

5. **Indicate which of the following statements apply to the jejunum and which to the ileum by placing ticks in the appropriate brackets.**

	In the jejunum	In the ileum
(a) The mucosa is thicker.	( )	( )
(b) The mucosa is more folded.	( )	( )
(c) The villi are shorter.	( )	( )
(d) The wall is thinner.	( )	( )

6. **Which of the statements on the right apply to the substances listed on the left?**

   (a) Iron.   (i) Absorbed into the lacteals.
   (b) Bile salts.   (ii) Absorbed in the duodenum.
   (c) Vitamin A.   (iii) Absorbed in the ileum.

ANSWERS TO TEST THREE

1.

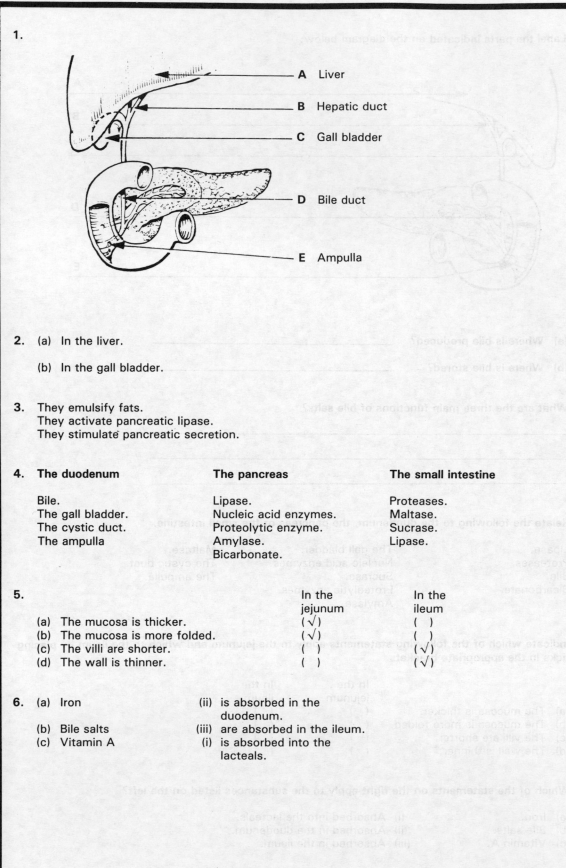

A Liver

B Hepatic duct

C Gall bladder

D Bile duct

E Ampulla

2. (a) In the liver.

(b) In the gall bladder.

3. They emulsify fats.
They activate pancreatic lipase.
They stimulate pancreatic secretion.

4. The duodenum	The pancreas	The small intestine
Bile.	Lipase.	Proteases.
The gall bladder.	Nucleic acid enzymes.	Maltase.
The cystic duct.	Proteolytic enzyme.	Sucrase.
The ampulla	Amylase.	Lipase.
	Bicarbonate.	

5.

	In the jejunum	In the ileum
(a) The mucosa is thicker.	(√)	( )
(b) The mucosa is more folded.	(√)	( )
(c) The villi are shorter.	( )	(√)
(d) The wall is thinner.	( )	(√)

6. (a) Iron   (ii) is absorbed in the duodenum.

(b) Bile salts   (iii) are absorbed in the ileum.

(c) Vitamin A   (i) is absorbed into the lacteals.

5. Movements in the alimentary tract

5.1. Introduction

The principal factor ensuring the passage of food down the alimentary tract is the action of the muscular layers of its wall.

There are two types of movement of the wall of the alimentary canal.

1. *Peristalsis*

This is a movement consisting of a **wave of contraction** preceded by a **wave of relaxation**, which propels the contents of the intestine along.

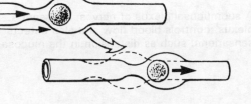

Peristaltic movements occur throughout the alimentary canal, but are most marked in the oesophagus after swallowing, in the pyloric antrum and upper small bowel during gastric emptying, and in the rectum during defecation.

2. *Segmentation*

This is a non-propulsive movement which mixes the chyme. The movement occurs in the large and small bowel, and more weakly, in the stomach.

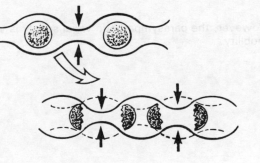

5.2. Co-ordination of movement

The smooth muscle of the bowel wall has an inherent tendency to contract rhythmically, even if it is completely isolated from any nerve supply.

The local *myogenic* activity is co-ordinated by a nerve network, the **myenteric plexus** (or Auerbach's plexus), which lies between the circular and longitudinal muscle layers.

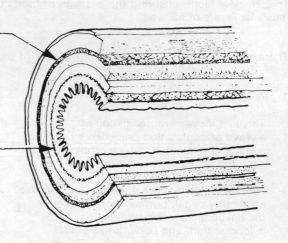

This plexus permits co-ordination of peristalsis and segmentation.

A submucosal plexus of nerves, **Meissner's plexus**, controls blood flow and also detects sensations, such as distension in the mucosa.

Although experiments suggest that sympathetic nerve impulses decrease motor activity of the bowel, and that parasympathetic nerve impulses increase motor activity, either of these autonomic nervous supplies can be cut in man, without any evident effect on the small or large bowel.

However, the parasympathetic vagal supply is very important in the control of stomach secretion and mobility.

5.3. Control of movements and secretions

The control of movements and secretions during a meal is governed by the parasympathetic vagus nerves and four hormones secreted by the antrum and upper small bowel.

The **vagus nerve** is the tenth cranial nerve. It arises from the brain stem, runs through the base of the skull and down the pharynx and oesophagus, to supply the stomach, liver and small bowel. Impulses in the vagus nerve cause a constant *basal secretion* of acid and pepsin by the gastric glands.

The sight, smell, anticipation or taste of food causes a reflex increase in the activity of the vagus nerve, increasing the secretion of acid and pepsin and increasing the movements of the muscular wall.

In most people the basal secretion is low but some people respond to mental stress with a constant high basal secretion of acid. This raised acidity can overcome the mucosal defences and lead to *duodenal ulceration*. The secretion of acid can be stopped by cutting the vagus nerve. This procedure must be combined with widening of the pylorus, since it causes the stomach to become flaccid and unable to empty properly.

Alternatively, the acid producing portion of the stomach can be removed.

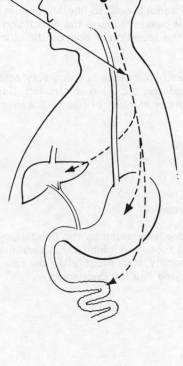

5.4. Gastrointestinal hormones

The gastrointestinal hormones are all polypeptides. They are secreted by the stomach and small intestine into the bloodstream, passing through the circulation to act on their target organ. Their target organs are all made more sensitive to their action by increased activity of the vagus nerve.

There are four important gastrointestinal hormones. Their properties overlap to a great extent.

Gastrin

Food in the pyloric antrum, especially meat, causes the antral G cells to liberate gastrin into the blood. It passes through the circulation and stimulates the secretion of acid by the stomach.

When the antral contents become very acid (ph ⩽ 2.5) gastrin secretion is inhibited. Gastrin increases the production of bile by the liver.

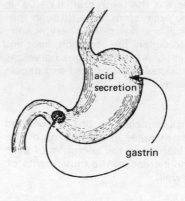

Enterogastrone

This hormone is secreted by the duodenum in response to food, and inhibits acid secretion and stomach motility, thus moderating the rate of entry of chyme into the duodenum.

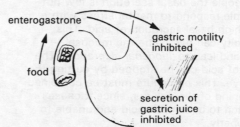

Cholecystokinin — pancreozymin (CCKPZ)

This substance was originally thought to be two hormones. It is released from the upper small bowel in response to food products. It causes pancreatic secretion, contraction of the gall bladder and relaxation of the ampullary sphincter.

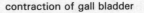

Secretin

Secretin is produced in response to acid in the duodenum and stimulates the pancreas to produce a watery alkaline fluid, the Brunner's glands to produce alkali, and the intestinal glands to produce mucus and enterokinase.

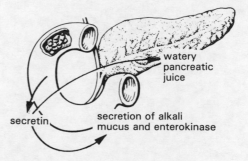

5.5. The large intestine

Chyme moves from the ileum into the caecum through the **ileo-caecal valve**, a fold of mucosa in the caecum which tends to prevent backflow of chyme.

The last 5 cm of the ileum acts as a sphincter. It is usually contracted. Filling of the stomach causes it to relax and the ileal contents enter into the caecum. This *gastro-colic reflex* is often associated with *mass movements*.

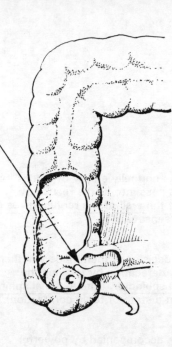

Mass movements are sudden rushes of peristalsis, beginning in mid-colon. They move the contents of the large intestine into the lower colon or even the rectum.

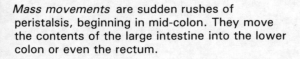
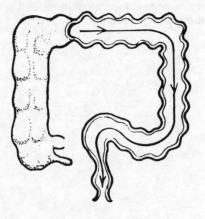

Segmental mixing movements also occur in the large intestine. They are co-ordinated by the myenteric plexus and bring the contents of the bowel into close contact with the mucosa, for the absorption of water and electrolytes.

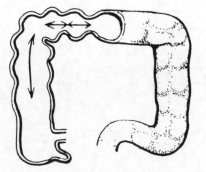

5.6. Defecation

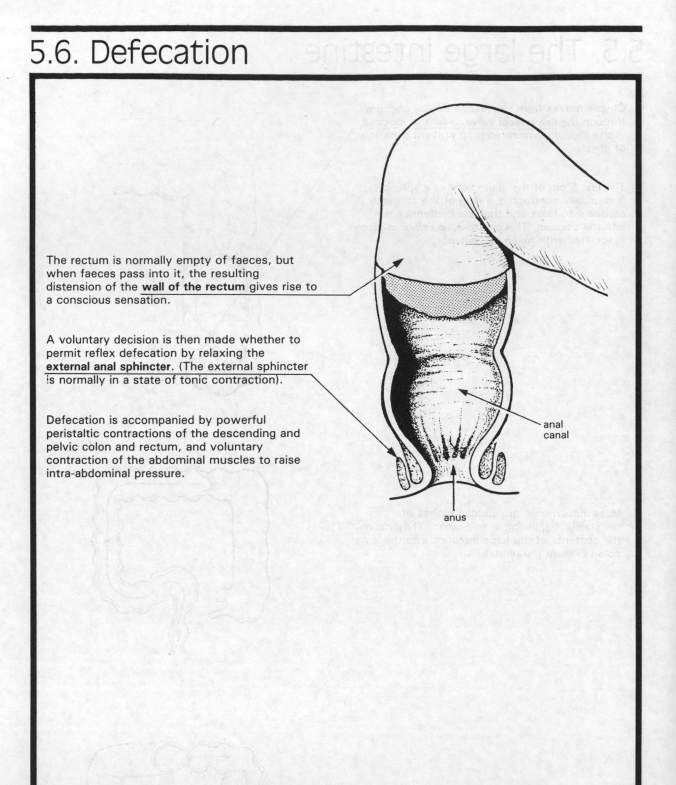

The rectum is normally empty of faeces, but when faeces pass into it, the resulting distension of the **wall of the rectum** gives rise to a conscious sensation.

A voluntary decision is then made whether to permit reflex defecation by relaxing the **external anal sphincter**. (The external sphincter is normally in a state of tonic contraction).

Defecation is accompanied by powerful peristaltic contractions of the descending and pelvic colon and rectum, and voluntary contraction of the abdominal muscles to raise intra-abdominal pressure.

anal canal

anus

TEST FOUR

1. **What is meant by the following?**

    (a) Peristalsis.
    (b) Mass movements.
    (c) Segmental mixing movements.

2. **Which of the statements on the right apply to the substances listed on the left?**

    (a) Gastrin.
    (b) Enterogastrone.
    (c) Cholecystokinin — pancreozymin.
    (d) Secretin.

    (i)   Causes contraction of the gall bladder.
    (ii) Provokes a watery alkaline pancreatic secretion.
    (iii) Reduces gastric emptying.
    (iv) Produced by the antrum.

3. **Indicate which of the statements below apply to segmentation and which to peristalsis by placing ticks in the appropriate brackets.**

	Segmentation	Peristalsis
(a) Occurs in the oesophagus.	( )	( )
(b) Propels food along the bowel.	( )	( )
(c) Mixes the chyme.	( )	( )
(d) Occurs in the large bowel.	( )	( )
(e) Depends on the myenteric plexus.	( )	( )

4. **Complete the following:**

    (a) Secretion of acid by the stomach and the production of bile by the liver are stimulated by
    (b) The rate of entry of chyme into the duodenum is moderated by
    (c) Pancreatic secretion, contraction of the gall bladder and relaxation of the ampullary sphincter are caused by
    (d) The pancreas is stimulated to produce an alkaline fluid, the Brunner's glands to produce alkali and the intestines to produce mucus and enterokinase by

ANSWERS TO TEST FOUR

1. (a) Peristalsis refers to the succeeding waves of muscular contraction and relaxation by which the food is propelled along the intestine.
   (b) Mass movement refers to the rushes of peristalsis beginning in mid-colon by which the contents of the large intestine are moved into the lower colon.
   (c) Segmental mixing movements refer to the movements in the large intestine which bring the contents of the bowel into contact with the mucosa and enable them to absorb water and electrolytes.

2. (a) Gastrin   (iv) is produced by the antrum.
   (b) Enterogastrone   (iii) reduces gastric emptying.
   (c) Cholecystokinin — pancreozymin   (i) causes contraction of the gall bladder.
   (d) Secretin   (ii) provokes a watery alkaline pancreatic secretion.

3.

	Segmentation	Peristalsis
(a) Occurs in the oesophagus.	( )	(√)
(b) Propels food along the bowel.	( )	(√)
(c) Mixes the chyme.	(√)	( )
(d) Occurs in the large bowel.	(√)	(√)
(e) Depends on the myenteric plexus.	(√)	(√)

4. (a) Secretion of acid by the stomach and the production of bile by the liver are stimulated by *gastrin*.

   (b) The rate of entry of chyme into the duodenum is moderated by *enterogastrone*.

   (c) Pancreatic secretion, contraction of the gall bladder and relaxation of the ampullary sphincter are caused by *cholecystokinin pancreozymin (CCKPZ)*.

   (d) The pancreas is stimulated to produce an alkaline fluid, the Brunner's glands to produce alkali and the intestines to produce mucus and enterokinase by *secretin*.

6. Transport — the portal system

6.1. Introduction

All the blood from the gut, from the lower oesophagus to the anus, flows into the **portal vein** and is carried to the **liver.**

The portal blood (1½ litres/min) carries all the water-soluble products of digestion (amino acids, fatty acids, glycerol, glucose and water-soluble vitamins) to the liver.

Fats, as droplet chylomicrons in the lacteals of the bowel wall, pass into the lymphatic ducts. These unite to form the **thoracic duct** which drains the lymph (10–20 ml/min) into the left subclavian vein, and thus into the circulation.

6. Transport—the portal system

6.2. The liver—anatomy

The liver occupies the upper part of the peritoneal cavity. It is protected by the lower ribs and lies directly under the **diaphragm**. Its shape is moulded by the organs surrounding it. It is covered by peritoneum and held in position by the inferior vena cava, the abdominal organs and peritoneal ligaments.

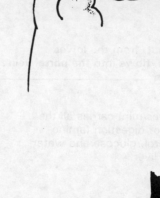

Blood from the digestive system enters the liver by the **portal vein**.

The liver is also supplied, with oxygenated arterial blood, by the **hepatic artery**.

All the blood after passing through the liver leaves via the **hepatic veins** which drain into the **inferior vena cava**.

Bile leaves the liver by the **hepatic ducts**, and empties into the **duodenum** through the **bile duct**.

The **gall bladder** is a side-arm of this system and serves as a reservoir for bile.

Hepatic artery

Portal vein

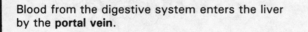

6.3. The liver—microscopic structure

The liver has a highly complex structure.

It is made up of *lobules*, which consist of sheets of liver cells arranged in radiating patterns around a central vein. The sheets of cells have blood spaces called **sinusoids** between them.

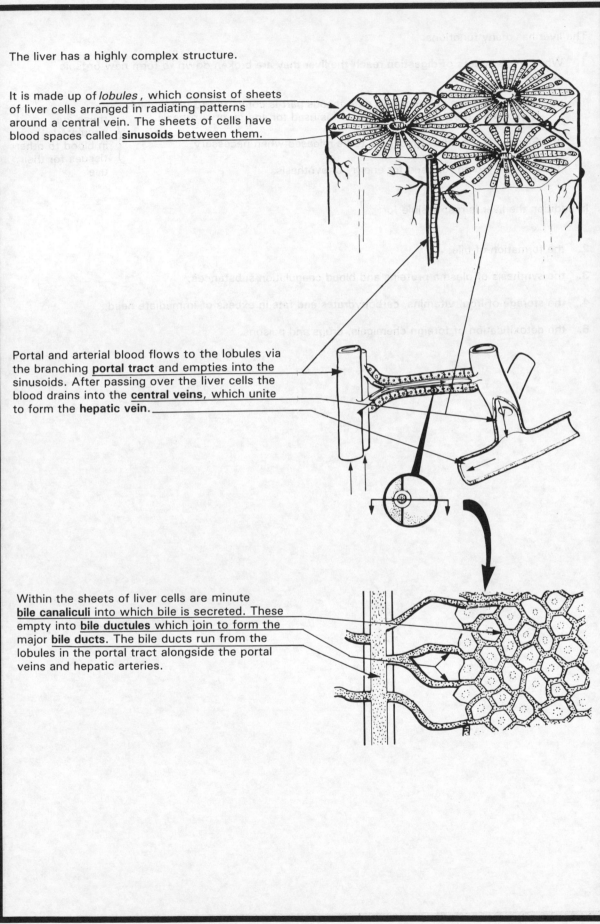

Portal and arterial blood flows to the lobules via the branching **portal tract** and empties into the sinusoids. After passing over the liver cells the blood drains into the **central veins**, which unite to form the **hepatic vein**.

Within the sheets of liver cells are minute **bile canaliculi** into which bile is secreted. These empty into **bile ductules** which join to form the major **bile ducts**. The bile ducts run from the lobules in the portal tract alongside the portal veins and hepatic arteries.

6.4. The liver—functions

The liver has many functions.

1. When the products of digestion reach the liver they are broken down to form new organic compounds.

   (a) Amino acids are split up. The nitrogenous part is converted to urea which is excreted; the non-nitrogenous part is used for energy or synthesis.

   (b) Glucose is stored as glycogen, to be released when necessary.

   (c) Fats are stored, or used for energy or synthesis.

   } these are circulated in blood to other tissues for their use

In addition the liver is responsible for:

2. the formation of bile;

3. the synthesis of plasma proteins and blood coagulation substances;

4. the storage of iron, vitamins, carbohydrates and fats in excess of immediate need;

5. the detoxification of foreign chemicals, drugs and poisons.

TEST FIVE

1. Which of the following organs have a blood supply draining into the portal vein?

    (a) The stomach.   (b) The liver.   (c) The pharynx.   (d) The caecum.   (e) The small bowel.

2. Which of the following blood vessels carry blood to the liver?

    (a) The hepatic vein.   (b) The hepatic artery.   (c) The portal vein.   (d) The inferior vena cava.

3. (a) What is the function of the portal vein?
    (b) What products of digestion are carried by portal blood?
    (c) What property of these products makes it possible for them to be so carried?

    _____

    _____

    _____

4. Allocate the following to *the entry of blood into the liver, the outflow of blood from the liver,* and *the outflow of bile from the liver*.

    Hepatic veins.
    Hepatic ducts.
    The inferior vena cava.
    The portal vein.
    The bile duct.
    The hepatic artery.

5. Which of the statements on the right apply to the items listed on the left?

    (a) Lacteal.   (i) Carries bile.
    (b) Portal vein.   (ii) Carries blood from the liver.
    (c) Inferior vena cava.   (iii) Carries milky lymph.
    (d) Hepatic duct.   (iv) Empties into the sinusoids.

ANSWERS TO TEST FIVE

1. (a) The stomach.   (d) The caecum.   (e) The small bowel.

2. (b) The hepatic artery.   (c) The portal vein.

3. (a) The portal vein carries all the blood from the gut to the liver.
   (b) Amino acids, fatty acids, glycerol, glucose and water soluble vitamins are carried by the portal blood.
   (c) Their water solubility makes it possible for these substances to be carried in the blood.

4. 1. *The entry of blood into the liver* is via the portal vein and the hepatic artery.
   2. *The outflow of blood from the liver* is via the hepatic veins and inferior vena cava.
   3. *The outflow of bile from the liver* is via the hepatic ducts and the bile duct.

5. (a) The lacteals   (iii) carry milky lymph.
   (b) The portal vein   (iv)   empties into the sinusoids.
   (c) The inferior vena cava   (ii)   carries blood from the liver.
   (d) The hepatic duct   (i)   carries bile.

POST TEST

1. **Indicate which of the names in the list below refer to the parts of the digestive system labelled on the diagram alongside, by placing the appropriate letters in the brackets.**

   1. The stomach.   ( )
   2. The oesophagus.   ( )
   3. The pancreas.   ( )
   4. The large intestine. ( )
   5. The gall bladder.   ( )
   6. The duodenum.   ( )
   7. The ileum.   ( )
   8. The liver.   ( )

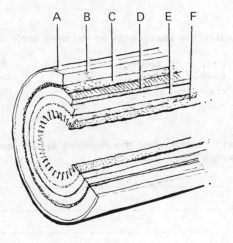

2. **Indicate which of the names in the list below refer to the various layers of the alimentary tract labelled on the diagram alongside, by placing the appropriate letters in the brackets.**

   1. The submucosa.   ( )
   2. The mucosa.   ( )
   3. The circular muscle. ( )
   4. The serosa.   ( )
   5. The longitudinal muscle. ( )
   6. The myenteric (or Auerbach's) plexus. ( )

3. **Which of the enzymes on the right act on the substances listed on the left?**

   (a) Starch.   (i) Amylase.
   (b) Maltose.   (ii) Trypsin.
   (c) Fats.   (iii) Maltase.
   (d) Proteins.   (iv) Lipase.

4. **Which of the enzymes on the right are liberated from the sites listed on the left?**

   (a) Small bowel.   (i) Ptyalin.
   (b) Pancreas.   (ii) Pepsin.
   (c) Parotid gland.   (iii) Trypsin.
   (d) Stomach.   (iv) Enterokinase.

POST TEST

5. Indicate which of the names in the list below refer to the parts of the digestive system labelled on the diagram alongside, by placing the appropriate letters in the brackets.

1. The cystic duct. ( )
2. The hepatic duct. ( )
3. The gall bladder. ( )
4. The ampulla. ( )
5. The pancreatic duct. ( )
6. The duodenum. ( )

6. (a) What kind of substances are the gastrointestinal hormones?
   (b) How do they reach their target organ?
   (c) Which nerve increases the sensitivity to the gastrointestinal hormones of the target organs?

   _____

   _____

   _____

7. Which of the statements on the right apply to the substances listed on the left?

   (a) Vitamin B_{12} and bile salts.   (i) Absorbed in the lower ileum.
   (b) Alcohol.   (ii) Absorbed in the stomach.
   (c) Water and electrolytes.   (iii) Absorbed in the duodenum.
   (d) Iron.   (iv) Absorbed in the large bowel.

8. What provision does the digestive system contain against the hydrochloric acid necessary for digestion?

   _____

   _____

   _____

   _____

9. Which of the statements on the right apply to the items listed on the left?

   (a) Micelles.   (i) Lymph spaces in the villi.
   (b) Lacteals.   (ii) Fat droplets synthesised in the villi.
   (c) Chylomicrons.   (iii) Complexes of fat digestion products and bile salts.
   (d) Triglycerides.   (iv) Water soluble.
   (e) Fatty acids.   (v) Main form of dietary fat.

POST TEST

10. Which of the following are functions of the liver?

(a) Storage of glycogen.
(b) Synthesis of bilirubin.
(c) Synthesis of blood coagulation substances.
(d) Synthesis of urea.
(e) Formation of bile.
(f) Storage of hormones.
(g) Detoxification of drugs.
(h) Formation of blood.

ANSWERS TO POST TEST

1. 1. The stomach. (C)
   2. The oesophagus. (A)
   3. The pancreas. (D)
   4. The large intestine. (F)
   5. The gall bladder. (B)
   6. The duodenum. (E)
   7. The ileum. (G)
   8. The liver. (H)

2. 1. The submucosa. (E)
   2. The mucosa. (F)
   3. The circular muscle. (D)
   4. Serosa. (A)
   5. The longitudinal muscle. (B)
   6. The myenteric (or Auerbach's) plexus. (C)

3. (a) Starch is acted on by   (i) amylase.
   (b) Maltose is acted on by   (iii) maltase.
   (c) Fats are acted on by   (iv) lipase.
   (d) Proteins are acted on by (ii) trypsin.

4. (a) The small bowel.   (iv) secretes enterokinase.
   (b) The pancreas   (iii) secretes trypsin.
   (c) The parotid gland (i) secretes ptyalin.
   (d) The stomach   (ii) secretes pepsin.

5. 1. The cystic duct. (A)
   2. The hepatic duct. (E)
   3. The gall bladder. (B)
   4. The ampulla. (D)
   5. The pancreatic duct. (F)
   6. The duodenum. (C)

6. (a) The gastrointestinal hormones are polypeptides.
   (b) Secreted by the bowel wall into the bloodstream, these hormones reach their target area via the circulation.
   (c) The vagus nerve increases the sensitivity of the target organs to the gastrointestinal hormones.

ANSWERS TO POST TEST

7. (a) Vitamin B$_{12}$ and bile salts   (i) are absorbed in the lower ileum.
  (b) Alcohol   (ii) is absorbed in the stomach.
  (c) Water and electrolytes   (iv) are absorbed in the large bowel.
  (d) Iron   (iii) is absorbed in the duodenum.

8.   (i) The gastric glands produce gastrin, an alkaline mucus.
  (ii) Mucous glands protect the stomach wall from its own juices.
  (iii) The pancreas secretes controlled amounts of bicarbonate to neutralise the chyme.
  (iv) Enterogastrone inhibits acid secretion thus moderating the rate of entry of chyme into the duodenum.
  (v) Secretin (like enterogastrone, a gastrointestinal hormone) stimulates the pancreas to produce an alkaline fluid, and the Brunner's glands to produce alkali.

9. (a) Micelles   (iii) are complexes of fat digestion products and bile salts.
  (b) Lacteals   (i) are lymph spaces in the villi.
  (c) Chylomicrons   (ii) are fat droplets synthesised in the villi.
  (d) Triglycerides   (v) are the main form of dietary fat.
  (e) Fatty acids   (iv) are water soluble.

10. The following are functions of the liver:

  (a) storage of glycogen;
  (c) synthesis of blood;
  (d) synthesis of urea;
  (e) formation of bile;
  (g) detoxification of drugs.